SPACE SCIENCE, EXPLORATION AND POLICIES

THE EARTH'S BEGINNING

SPACE SCIENCE, EXPLORATION AND POLICIES

Additional books and e-books in this series can be found on Nova's website under the Series tab.

SPACE SCIENCE, EXPLORATION AND POLICIES

THE EARTH'S BEGINNING

ROBERT S. BALL

snova
New York

Library of Congress Cataloging-in-Publication Data

ISBN: 978-1-53615-743-7

Published by Nova Science Publishers, Inc. † New York

CONTENTS

LIST OF ILLUSTRATIONS

FOREWORD

Since these lectures were delivered in the Royal Institution of Great Britain there has been much advance in our knowledge of astronomy. The simultaneous advance in other sciences allied with astronomy has been, perhaps, even more remarkable. I am glad to avail myself of the opportunity afforded by a new issue of "The Earth's Beginning" to draw attention to certain recent developments of science which relate in a very striking way to the subject of this volume, namely, the famous Nebular Theory of the origin of the solar system. It appears to me that these recent developments tend to reduce greatly, even if they do not altogether remove, the chief outstanding difficulty which has hitherto retarded the acceptance of the Nebular Theory.

I have explained in Chapter 6. those views of Helmholtz which have for so long provided the received explanation of the maintenance of solar heat. Calculation shows that if the sun's heat has been maintained by the contraction of the primæval nebula—and this was the supposition of Helmholtz—the orb of day cannot have radiated with its present intensity for a period much longer than twenty million years.

But from the evidence of geology it must now be admitted that the existence of our earth, indeed even that part of its existence during which it has been the abode of life, has endured for a period far in excess of that which this calculation would allow. It therefore seems to follow that the theory of Helmholtz does not provide an adequate explanation of such an amazing phenomenon as the continuance of a sufficient supply of sunbeams throughout the vast periods demanded by geological phenomena.

There is another entirely different line of reasoning by which Professor John Joly has recently taught us the immense antiquity of our earth. His argument is based upon an estimate of the time that must have elapsed since the waters of the ocean, which had previously been sustained in the great vapours of the atmosphere, were deposited in the ocean beds. When the earth had become sufficiently cool to permit of the vapours now forming the ocean passing from the gaseous to the liquid form, the oceans descended from the heavens above to the earth beneath in the form of *fresh* water. In the lapse of subsequent ages the sea has become salt because ordinary river water, which always contains some small quantity of salt in solution, is continually bearing salt down to the sea. No doubt water is constantly being abstracted from the sea by evaporation, but only *fresh* water is thus removed, so in this cycle of change the salt in the sea must be gradually accumulating. Thus, day by day, though no doubt extremely slowly, the sea has been growing more and more salt.

Professor Joly has made an estimate of the quantity of salt daily added to the sea by all the rivers of the globe. He has also made an estimate of the total quantity of salt which is at present contained in the sea. He has thus the means of forming an estimate of the number of years necessary for the sea to have become converted from its primæval freshness to its present saltness. His result is not a little astonishing. The saltness of the sea could not be accounted for unless the rivers had been running into the sea for at least a hundred million years. This period is five times as long as the total period during which the sun could have been shining if the Helmholtzian view were correct.

Of course, there are many elements of uncertainty in such a calculation. We have assumed that the total flow of the rivers is practically constant, and that our estimate fairly represents the average salinity of river water. We have also made a large assumption in supposing that we have accurately estimated the total volume of salt in the oceans. But taken in conjunction with the geological evidence already referred to, taken in conjunction with the immense periods of time that have been required for the evolution of life on the globe by the process of natural selection, the conclusion arrived at is inevitable. It seems impossible to doubt that the sun must have been shining and that our solar system must have existed in practically the same form as it is at present for periods enormously greater than would have been possible if the heat of the sun had been sustained by the solar contraction only.

The difficulty here indicated has been not unjustly considered the most serious difficulty with which the development of modern physical and astronomical science has been confronted. The time during which the sun must have lasted, according to the received explanation of the source of its heat and the time during which the sun has actually lasted, as shown by the facts of geology, present a wide discrepancy. Science demands that some reconciliation must be effected, yet how is that to be accomplished? There is only one possible solution of the problem. It is obvious that there must have been some vast reserve of heat in the sun in comparison with which the quantity of heat yielded by the contraction may be deemed insignificant. Until this new source of solar energy had been discovered, our knowledge of the physics of the solar system lay under a reproach, which it was the bounden duty of men of science to endeavour to remove.

During the last few years lines of research carried on in various directions have, in a most unexpected manner, thrown much light on the origin of the sun's heat, and, indeed, we may now say that the great difficulty which has for so long troubled us no longer exists in a serious form.

Recent discoveries show that matter possesses stores of energy which, if not actually boundless, are enormously in excess of what had been previously deemed possible. These stores of energy are available for supplying the heat of the sun, and it is easy to show that they are amply sufficient to furnish the necessary sunbeams for even the longest periods during which the claims of geology maintain that the sun must have been shining.

The researches of Professor Sir J. J. Thomson have shown how corpuscles of matter are sometimes moving with velocities enormously greater than those of any celestial body with which astronomy had made us acquainted. The case of high corpuscular velocity which is most generally known is that presented by radium, the particles from which are being continually shot forth in myriads. It is quite true that each of these corpuscles is excessively small, and it may be useful to give the following illustration bearing on the subject. Think of a number represented by unity followed by eighteen cyphers, or more concisely as 10^{18}, and think of a line a kilometre long. If that line were divided into 10^{18} parts, each of those parts would represent the diameter of a corpuscle of radium. If that line were multiplied by 10^{18}, the result would be a line so long that a ray of light would require a period of no less than 100,000 years to pass from one end to the other.

These corpuscles of radium are, no doubt, excessively small, but the velocity with which they are moving is comparable with the velocity of light. When a material object is moving with a velocity of that magnitude the energy it contains in virtue of that velocity is indeed startling. A very small grain of sand would, if moving with the velocity of light, contain, in virtue of that motion, the equivalent of more heat than could be produced by the combustion of a ton of the best coal. The late Dr. W. E. Wilson showed that if an excessively minute percentage of radium should be found to exist in the sun, it would completely account for the sustentation of the solar heat, and the Hon. R. Strutt has shown that the minute quantities of radium which he has proved to exist in terrestrial rocks would enormously protract the earth's cooling. These discoveries have, in fact, completely changed the outlook on the problem of the sun's heat, and, though no doubt much has yet to be done before the whole subject is cleared up, the great difficulty may be regarded as vanquished. Thus, the discovery of radium, and the wonderful phenomena associated therewith, has pointed out a possible escape from one of the gravest difficulties in science.

The most notable fact which emerges from the modern study of the structure of the heavens is the ever-increasing significance and importance of the spiral nebulæ. The following pages will have failed in their object if they have not succeeded in emphasising the fact that the spiral nebula is, next to a fixed star itself, the most characteristic type of object in the material universe. With every increase in the power of the telescope, and with every development of the application of photography to celestial portraiture, the importance of the spiral structure in nebulæ becomes of ever-increasing interest.

But I revert to this subject here for the purpose of taking notice of a suggestive paper by Mr. C. Easton in the "Astrophysical Journal," Vol. XII., No. 2, September, 1900, entitled "A New Theory of the Milky Way." This paper advances the striking view that the Milky Way is itself a spiral nebula, and certainly the considerations adduced by Mr. Easton seem to justify his remarkable conclusion.

It is first to be noticed that the Milky Way extends as an irregular band completely round the heavens, and that it follows very nearly the course of a great circle. The curious convolutions of the Milky Way, the varying star densities of its different parts, would, as shown by Mr. Easton, be completely accounted for if the Milky Way were a mighty spiral. We view the ordinary celestial spirals from the outside at an immense distance in space. We view the Milky Way from a position within the circuit of the nebula. It has, however, been shown by Mr. Easton that the centre of the Spiral Nebula is not exactly at the sun. The centre of the Milky Way is near that superb region of the galaxy which lies in Cygnus.

Thus, the significance of the spiral structure in the universe becomes greatly enhanced. The spirals abound in every part of the heavens; they are placed in every conceivable position and in every possible plane; they have every range in size from comparatively small objects, whose destiny is to evolve into a system like our solar system, up to stupendous objects which include a myriad of such systems. There is now the further interest that as the sun and the solar system are included within the Milky Way, and as the Milky Way is a spiral, this earth of ours is itself at this moment a constituent part of a great spiral.

Finally, I would say that, so far as I have been able to understand the subject, it appears to me that every advance in our knowledge of the heavens tends more and more to support the grand outlines of the Nebular Theory as imagined by Kant and Laplace.

R. S. B.

May 1, 1909.

Figure 1. An English Sunset Tinged by Krakatoa. (From a Drawing made at Chelsea at 4.40 p.m. on Nov. 26th, 1883, by Mr. W. Ascroft.)

PREFACE*

The Earth's Beginning are lectures which were delivered in the Royal Institution of Great Britain. It considers the majestic subject of the evolution of the solar system of which our earth forms a part. The nebular theory discloses the beginning of this earth itself. It shows how the foundations of this solid earth have been laid, and how it is that we have land to tread on and air to breathe. But the subject has a scope far wider than merely in its relation to our earth.

---

* This is an edited, augmented and reformatted version of "The Earth's Beginning," by Robert S. Ball, originally published by Cassell and Company, Limited., dated 1901.

Chapter 1

Introduction

I try in these lectures to give some account of an exceptionally great subject—a subject, I ought rather to say, of sublime magnificence. It may, I believe, be affirmed without exaggeration that the theme which is to occupy our attention represents the most daring height to which the human intellect has ever ventured to soar in its efforts to understand the great operations of Nature. The earth's beginning relates to phenomena of such magnitude and importance that the temporary concerns which usually engage our thoughts must be forgotten in its presence. Our personal affairs, the affairs of the nation, and of the empire—indeed, of all nations and of all empires—nay, even all human affairs, past, present, and to come, shrink into utter insignificance when we are to consider the majestic subject of the evolution of that solar system of which our earth forms a part. We shall obtain a glimpse of what that evolution has been in the mighty chapter of the book of Nature on which we are now to enter.

The nebular theory discloses the beginning of this earth itself. It points out the marvellous process by which from original chaos the firm globe on which we stand was gradually evolved. It shows how the foundations of this solid earth have been laid, and how it is that we have land to tread on and air to breathe. But the subject has a scope far wider than merely in its relation to our earth. The nebular theory accounts for the beginning of that great and glorious orb the sun, which presides over the system of revolving planets, guides them in their paths, illuminates them with its light, and stimulates the activities of their inhabitants with its genial warmth. The nebular theory explains how it comes about that the sun still continues in these latter days to shine with the brilliance and warmth that it had throughout the past ages of human history and the vastly greater periods of geological time. Then, as another supreme achievement, it discloses the origin of the planets which accompany the sun, and shows how they have come to run their mighty courses; and it tells us how revolving satellites have been associated with the planets. The nebular theory has, indeed, a remarkable relation to all objects belonging to that wonderful scheme which we call the solar system.

It should also be noticed that the nebular theory often brings facts of the most diverse character into striking apposition. As it accounts for the continued maintenance of the solar radiation, so it also accounts for that beneficent rotation by which each continent, after the enjoyment of a day under the invigorating rays of the sun, passes in due alternation into the repose of night. The nebular theory is ready with an explanation of the marvellous structure revealed in the rings of Saturn, and it shows at the same time how the volcanoes of the moon acquired their past phenomenal activity, and why, after ages of activity, they have now at last become extinct. With equal versatility the nebular theory will explain why a collier experiences increasing heat as he descends the coalpit, and why the planet Jupiter is marked with those belts which have so much interest for the astronomer. The nebular theory offers an immediate explanation of the earthquake which wrought such awful destruction at Lisbon, while it also points out the cause of that healing warmth of the waters at Bath. Above all, the nebular theory explains that peerless discovery of cosmical chemistry which declares that those particular elements of which the sun is composed are no other than the elements which form the earth beneath our feet.

When a doctrine of such transcendent importance is proposed for our acceptance, it is fitting that we should look, in the first instance, to the source from which the doctrine has emanated. It would already have made good its claim to most careful hearing, though not perhaps to necessary acceptance, if it came to us bearing credentials which prove it to be the outcome of the thought and research of one endowed with the highest order of intellect. If the nebular theory had been propounded by only a single great leader of thought, the sublimity of the subject with which it deals would have compelled the attention of those who love to study the book of Nature. If it had appeared that a second investigator, also famous for the loftiest intellectual achievement, had given to the nebular theory the sanction of his name, a very much stronger claim for its consideration would at once have been established. If it should further appear that yet a third philosopher, a man who was also an intellectual giant, had been conducted to somewhat similar conclusions, we should admit, I need hardly say, that the argument had been presented with still further force. It may also be observed that there might even be certain conditions in the work of the three philosophers which would make for additional strength in the cause advocated; if it should be found that each of the great men of science had arrived at the same conclusion irrespective of the others, and, indeed, in total ignorance of the line of thought which his illustrious compeers were pursuing, this would, of course, be in itself a corroboration. If, finally, the methods of research adopted by these investigators had been wholly different, although converging to the establishment of the theory, then even the most sceptical might be disposed to concede the startling claim which the theory made upon his reason and his imagination.

All the conditions that I have assumed have been fulfilled in the presentation of the nebular theory to the scientific world. It would not be possible to point to three names more eminent in their respective branches of knowledge than those of Kant, Laplace, and William Herschel. Kant occupies a unique position by the profundity and breadth of his philosophical studies; Laplace applied the great discoveries of Newton to the investigation of the movements of the heavenly bodies, publishing the results in his immortal work, *Mécanique Céleste*; Herschel has been the greatest and the most original observer of the heavens since the telescope was invented. It is not a little remarkable that the great philosopher from his profound meditation, the great mathematician from a life devoted to calculations about the laws of Nature, the great observer from sounding the depths of the firmament, should each in the pursuit of his own line of work have been led to believe that the grand course of Nature is essentially expressed by the nebular theory. There have been differences of detail in the three theories; indeed, there have been differences in points which are by no means unimportant. This was unavoidable in the case of workers along lines so distinct, and of a subject where many of the elements were still unknown, as indeed many are still. Even at the present day no man can give a complete account of what has happened in the great evolution. But the monumental fact remains that these three most sagacious men of science, whose lives were devoted to the pursuit of knowledge, each approaching the subject from his own direction, each pursuing his course in ignorance of what the others were doing, were substantially led to the same result. The progress of knowledge since the time when these great men lived has confirmed, in ways which we shall endeavour to set forth, the sublime doctrine to which their genius had conducted them.

Immanuel Kant, whose grandfather was a Scotsman, was born in 1724 at Königsberg, where his life was spent as a professor in the University, and where he died in 1804. In the announcement of the application of the principle of evolution to the solar system, Laplace was preceded by this great German philosopher. The profound thinker who expounded the famous doctrine of time and space did not disdain to allow his attention to be also occupied with things more material than the subtleties of metaphysical investigation. As a natural philosopher Kant was much in advance of his time. His speculations on questions relating to the operations in progress in the material universe are in remarkable conformity with what is now accepted as the result of modern investigation. Kant outlined with a firmness inspired by genius that nebular theory to which Laplace subsequently and independently gave a more definite form, and which now bears his name.

Kant's famous work with which we are now concerned appeared in 1755.[1] In it he laid down the immortal principle of the nebular theory.

Figure 2. Immanuel Kant. *(From an old Print.)*

---

[1] We are now fortunately able to refer the English reader to the work of Professor W. Hastie, D.D., entitled "Kant's Cosmogony," Glasgow, 1900. Kant's most interesting career is charmingly described in De Quincey's "Last Days of Immanuel Kant."

The greatness of this book is acknowledged by all who have read it, and notwithstanding that the progress of knowledge has made it obvious that many of the statements it contains must now receive modification, Kant's work contains the essential principle affirming that the earth, the sun, the planets, and all the bodies now forming the solar system did really originate from a vast contracting nebula. In later years Kant's attention was diverted from these physical questions to that profound system of philosophy with which his name is chiefly associated. The nebular theory is therefore to be regarded as incidental to Kant's great lifework rather than as forming a very large and important part of it.

At the close of the last century, while France was in the throes of the Revolution, a school of French mathematicians was engaged in the accomplishment of a task which marked an epoch in the history of human thought. Foremost among the mathematicians who devoted their energies to the discussion of the great problems of the universe was the illustrious Laplace. As a personal friend of Napoleon, Laplace received marked distinction from the Emperor, who was himself enough of a mathematician to be able to estimate at their true value the magnificent results to which Laplace was conducted.

It was at the commencement of Kant's career, and before his great lifework in metaphysics was undertaken, that he was led to his nebular theory of the solar system. In the case of Laplace, on the other hand, the nebular theory was not advanced until the close of the great work of his life. The *Mécanique Céleste* had been written, and the fame of its author had been established for all time; and then in a few pages of a subsequent volume, called the *Système du Monde*, he laid down his famous nebular theory. In that small space he gave a wonderful outline of the history of the solar system. He had not read that history in any books or manuscripts; he had not learned it from any ancient inscriptions; he had taken it direct from the great book of Nature.

Influenced by the caution so characteristic of one whose life had been devoted entirely to the pursuit of the most accurate of all the sciences, Laplace accompanied his announcement of the nebular theory with becoming words of warning. The great philosopher pointed out that there are two methods of discovering the truths of astronomy. Some truths may be discovered by observing the heavenly bodies with telescopes, by measuring with every care their dimensions and their positions, and by following their movements with assiduous watchfulness. But there is another totally different method which has enabled many remarkable discoveries to be made in astronomy; for discoveries may be made by mathematical calculations which have as their basis the numerical facts obtained by actual observation. This mathematical method often yields results far more profound than any which can be obtained by the astronomer's telescope. The pen of the mathematician is indeed an instrument which sometimes anticipates revelations that are subsequently confirmed by actual observation. It is an instrument which frequently performs the highly useful task of checking the deductions that might too hastily be drawn from telescopic observations. It is an instrument the scope of whose discoveries embraces regions immeasurably beyond the reach of the greatest telescope. The pen of the mathematician can give us information as to events which took place long before telescopes came into existence—nay, even unnumbered ages prior to the advent of man on this earth.

Laplace was careful to say that the nebular theory which he sketched must necessarily be judged by a standard different from that which we apply to astronomical truths revealed by telescopic observation or ascertained by actual calculation. The nebular theory, said the great French mathematician, has to be received with caution, inasmuch as from the nature of the case it cannot be verified by observation, nor does it admit of proof possessing mathematical certainty.

A large part of these lectures will be devoted to the evidence bearing upon this famous doctrine. Let it suffice here to remark that the quantity of evidence now available is vastly greater than it was a hundred years ago, and furthermore, that there are lines of evidence which can now be followed which were wholly undreamt of in the days of Kant and Laplace. The particular canons laid down by Laplace, to which we have just referred, are perhaps not regarded as so absolutely binding in modern days. If we were to reject belief in everything which cannot be proved either by the testimony of actual eye-witnesses or by strict mathematical deductions, it would, I fear, fare badly with not a few great departments of modern science. It will not be necessary to do more at present than just to mention, in illustration of this, the great doctrine of the evolution of life, which accounts for the existing races of plants and animals, including even man himself. I need hardly say that the Darwinian theory, which claims that man has come by lineal descent from animals of a lower type, admits of no proof by mathematics; it receives assuredly no direct testimony from eye witnesses; and yet the fact that man has so descended is, I suppose, now almost universally admitted.

In the case of the great German philosopher, as well as in the case of the great French mathematician, the enunciation and the promulgation of their nebular theories were merely incidental to the important scientific undertakings with which their respective lives were mainly occupied. The relation of the nebular theory to the main lifework of the third philosopher I have named, has been somewhat different. When William Herschel constructed the telescopes with which, in conjunction with his illustrious sister, he conducted his long night-watches, he discovered thousands of new nebulæ; he may, in fact, be said to have created nebular astronomy as we now know it. Ever meditating on the objects which his telescopes brought to light, ever striving to sound the mysteries of the universe, Herschel perceived that between a nebula which was merely a diffused stain of light on the sky, and an object which was hardly distinguishable from a star with a slight haze around it, every intermediate grade could be found. In this way he was led to the splendid discovery which announced the gradual transformation of nebulæ into stars. We have already noted how the profound mathematician was conducted to a view of the origin of the solar system which was substantially identical with that which had been arrived at by the consummate metaphysician. The interest is greatly increased when we find that similar conclusions were drawn independently from the telescopic work of the most diligent and most famous astronomical observer who has ever lived. Not from abstract speculation like Kant, not from mathematical suggestion like Laplace, but from accurate and laborious study of the heavens was the great William Herschel led to the conception of the nebular theory of evolution.

That three different men of science, approaching the study of perhaps the greatest problem which Nature offers us from points of view so fundamentally different, should have been led substantially to the same result, is a remarkable incident in the history of knowledge. Surely the theory introduced under such auspices and sustained by such a weight of testimony has the very strongest claim on our attention and respect.

In the discussion on which we are about to enter in these lectures we must often be prepared to make a special effort of the imagination to help us to realise how greatly the scale of the operations on which the attention is fixed transcends that of the phenomena with which our ordinary affairs are concerned. Our eyes can explore a region of space which, however vast, must still be only infinitesimal in comparison with the extent of space itself. Notwithstanding all that telescopes can do for us, our knowledge of the universe must be necessarily restricted to a mere speck in space, a speck which bears to the whole of space a ratio less—we might perhaps say infinitely less—than that which the area of a single daisy bears to the area of the continent where that daisy blooms. But we need not repine at this limitation; a whole life devoted to the study of a daisy would not be long enough to explore all the mysteries of its life. In like manner the duration of the human race would not be long enough to explore adequately even that small part of space which is submitted for our examination.

But it is not merely the necessary limits of our senses which restrict our opportunities for the study of the great phenomena of the universe. Man's life is too short for the purpose. That our days are but a span is the commonplace of the preacher. But it is a commonplace specially brought home to us in the study of the nebular theory. A man of fourscore will allude to his life as a long one, and no doubt it may be considered long in relation to the ordinary affairs of our abode on earth; but what is a period of eighty years in the history of the formation of a solar system in the great laboratory of the universe? Such a period then seems to be but a trifle—it is nothing. Eighty years may be long enough to witness the growth of children and grandchildren; but it is too short for a single heartbeat in the great life of Nature. Even the longest lifetime is far too brief to witness a perceptible advance in the grand transformation. The periods of time demanded in the great evolution shadowed forth by the nebular theory utterly transcend our ordinary notions of chronology. The dates at which supreme events occurred in the celestial evolution are immeasurably more remote than any other dates which we are ever called upon to consider in other departments of science. The time of the story on which we are to be engaged is earlier, far earlier, than any date we have ever learned at school, or have ever forgotten since. The incidents of that period took place long before any date was written in figures—earlier than any of those very ancient dates which the geologists indicate not by figures indeed, but by creatures whose remains imbedded in the rocks suffice to give a character to the period referred to. The geologist will specify one epoch as that in which the fossilized bone of some huge extinct reptile was part of a living animal; he may specify another by the statement that the shell of some beautiful ammonite was then inhabited by a living form which swam in the warm primæval seas. The date of our story has at least this much certainty: that it is prior—immeasurably prior—to the time when that marvellous thing which we call life first came into being.

Voltaire has an instructive fable which I cannot resist repeating. It will serve, at all events, to bring before us the way in which the lapse of time ought to be regarded by one who desires to view the great operations of Nature in their proper proportions. He tells how an inhabitant of the star Sirius went forth on a voyage of exploration through the remote depths of space. In the course of his travels he visited many other worlds, and at length reached Saturn, that majestic orb, which revolved upon the frontier of the solar system, as then known. Alighting on the ringed globe for rest and investigation, the Sirian wanderer, in quest of knowledge, was successful in obtaining an interview with a stately inhabitant of Saturn who enjoyed the reputation of exceptional learning and wisdom. The Sirian hoped to have some improving conversation with this sage who dwelt on a globe so utterly unlike his own, and who had such opportunities of studying the majestic processes of Nature in remote parts of the universe.

He thought perhaps they might be able to compare instructive notes about the constitution of the suns and systems in their respective neighbourhoods. The visitor accordingly prattled away gaily. He opened all his little store of knowledge about the Milky Way, about the Great Bear, and about the great Nebula in Orion; and then pausing, he asked what the Saturnian had to communicate in reply. But the philosopher remained silent. Eagerly pressed to make some response, the grave student who dwelt on the frontier globe at last said in effect: "Sirian, I can tell you but little of Nature. I can tell you indeed nothing that is really worthy of the great theme which Nature proposes; for the grand operations of Nature are very slow; they are so slow that the great transformations in progress around us would have to be watched for a very long time before they could be properly understood. To observe Nature so as to perceive what is really happening, it would be necessary to have a long life; but the lives of the inhabitants of Saturn are not long; none of us ever lives more than fifteen thousand years."

Change is the order of Nature. Many changes no doubt take place rapidly, but the great changes by which the system has been wrought into its present form, those profound changes which have produced results of the greatest magnificence in celestial architecture are extremely slow. We should make a huge mistake if we imagined that changes—even immense changes—are not in progress, merely because our brief day is too short a period wherein to perceive them.

On the village green stands an oak-tree, a veteran which some say dates from the time of William the Conqueror, but which all agree must certainly have been a magnificent piece of timber in the days of Queen Elizabeth. The children play under that tree just as their parents and their grandparents did before them. A year, a few years, even a lifetime, may show no appreciable changes in a tree of such age and stature. Its girth does not perceptibly increase in such a period. But suppose that a butterfly whose life lasts but a day or two were to pass his little span in and about this venerable oak. He would not be able to perceive any changes in the tree during the insignificant period over which his little life extended.

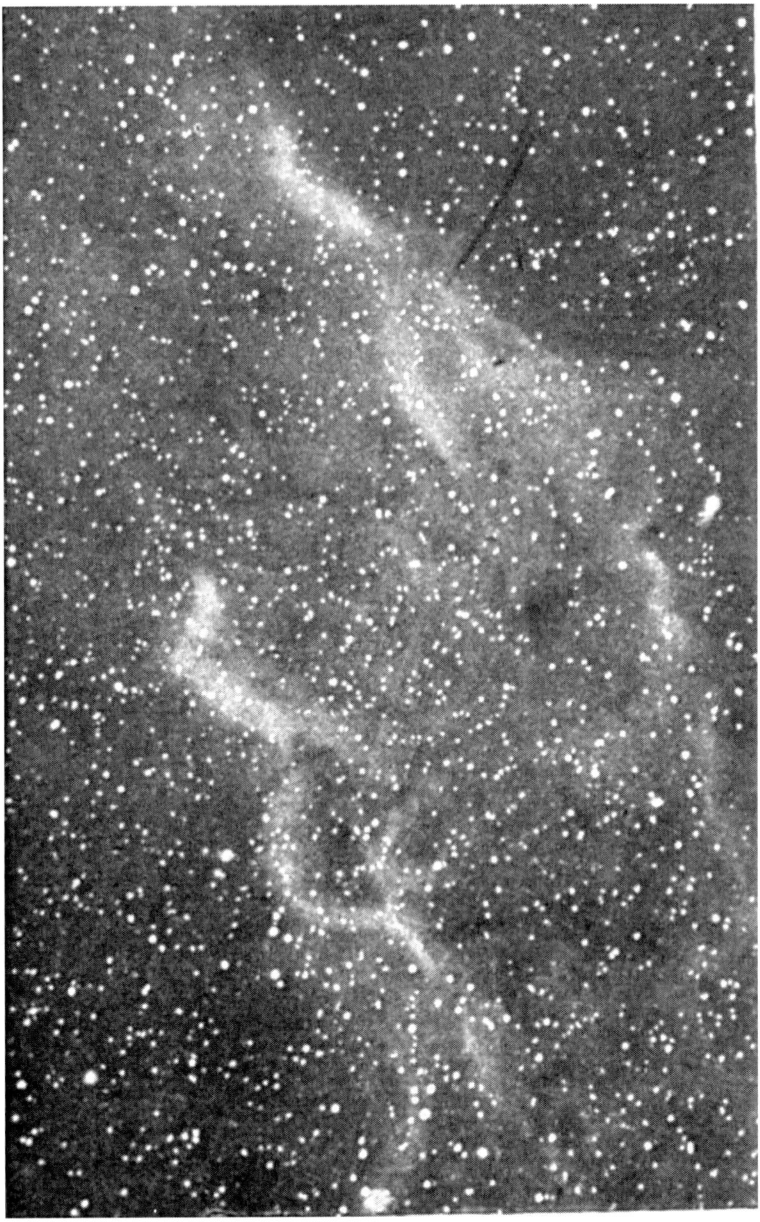

Figure 3. A Faint Diffused Nebulosity (n.g.c. 1499; in Perseus). (*Photographed by Dr. Isaac Roberts, F.R.S.*)

Not alone the mighty trunk and the branches, but even the very foliage itself would seem essentially the same in the minutes of the butterfly's extreme old age as they did in the time of his life's meridian or at the earliest moment of his youth. To the observations of a spectator who viewed it under such ephemeral conditions the oak-tree would appear steadfast, and might incautiously be deemed eternal. If the butterfly could reflect on the subject, he might perhaps argue that there could not be any change in progress in the oak-tree, because although he had observed it carefully all his life he could not detect any certain alteration. He might therefore not improbably draw the preposterous conclusion that the oak-tree must always have been just as large and just as green as he had invariably known it; and he might also infer that just as the oak-tree is now, so will it remain for all time. In our study of the heavens we must strive to avoid inferences so utterly fallacious as these which I have here tried to illustrate. Let it be granted that to our superficial view the sun and the moon, the stars and the constellations present features which appear to us as eternal as the bole of the oak seemed to the butterfly. But though the sun may seem to us always of the same size and always of the same lustre, it would be quite wrong to infer that the lustre and size of the sun are in truth unchanging. The sun is no more unchanging than the oak-tree is eternal. The sun and the earth, no less than the other bodies of the universe, are in process of a transformation no less astonishing than that wonderful transformation which in the course of centuries develops an acorn into the giant of the forest. We could not indeed with propriety apply to the great transformation of the sun the particular word growth; the character of the solar transformation cannot be so described. The oak-tree, of course, enlarges with its years, while the sun, on the other hand, is becoming smaller. The resemblance between the sun and the oak-tree extends no further than that a transformation is taking place in each. The rate at which each transformation is effected is but slow; the growth of the oak is too slow to be perceived in a day or two; the contraction of the sun is too slow to be appreciable within the centuries of human history.

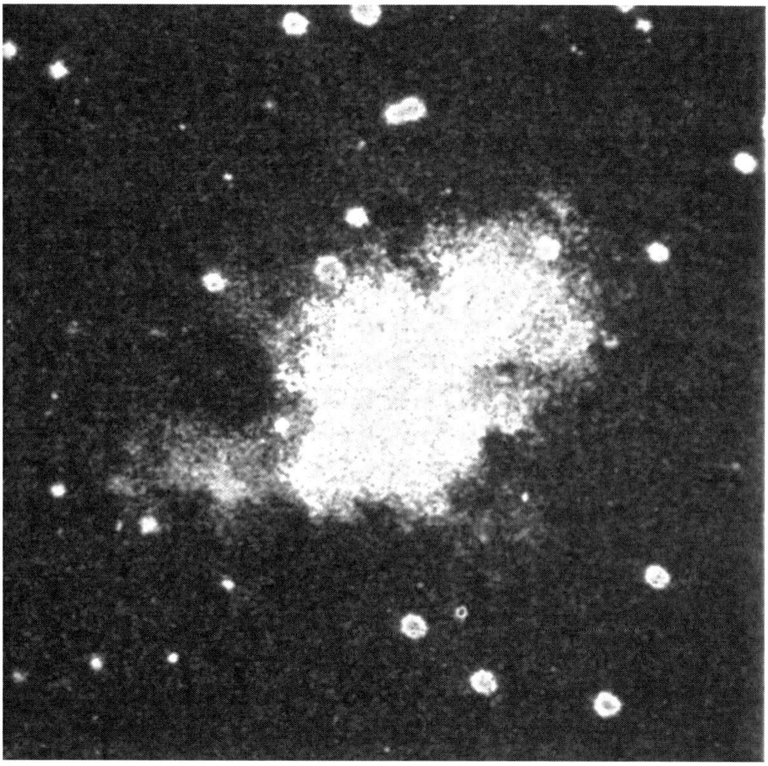

Figure 4. The Crab Nebula (n.g.c. 1952; in Taurus). (*Photographed by Dr. Isaac Roberts, F.R.S*)

Whatever the butterfly's observation might have suggested with regard to the eternity of the oak, we know there was a time when that oak-tree was not, and we know that a time will come when that oak-tree will no longer be. In like manner we know there was a time when the solar system was utterly different from the solar system as we see it now; and we know that a time will come when the solar system will be utterly different from that which we see at present. The mightiest changes are most certainly in progress around us. We must not deem them non-existent, merely because they elude our scrutiny, for our senses may not be quick enough to perceive the small extent of some of these changes within our limited period of observation. The intellect in such a case confers on man a power of surveying Nature with a penetration immeasurably beyond that afforded by his organs of sense.

That the great oak-tree which has lived for centuries sprang from an acorn no one can doubt; but what is the evidence on which we believe this to have been the origin of a veteran of the forest when history and tradition are both silent? In the absence of authentic documents to trace the growth of that oak-tree from the beginning, how do we know that it sprouted from an acorn? The only reason we have for believing that the oak-tree has gone through this remarkable development is deduced from the observation of other oak-trees. We know the acorn that has just sprouted; we know the young sapling as thick as a walking stick; we know the vigorous young tree as stout as a man's arm or as his body; we know the tree when it first approaches the dignity of being called timber; we can therefore observe different trees grade by grade in a continuous succession from the acorn to the monarch of five centuries. No one doubts for a moment that the growth as witnessed in the stages exhibited by several different trees, gives a substantially accurate picture of the development of any individual tree. Such is the nature of one of the arguments which we apply to the great problem before us. We are to study what the solar system has been in the course of its history by the stages which we witness at the present moment in the evolution of other systems throughout the universe. We cannot indeed read the history in time, but we can read it in space.

The mighty transformation through which the solar system has passed, and is even now at this moment passing, cannot be actually beheld by us poor creatures of a day. It might perhaps be surveyed by beings whose pulses counted centuries, as our pulses count seconds, by beings whose minutes lasted longer than the dynasties of human history, by beings to whom a year was comparable with the period since the earth was young, and since life began to move in the waters.

May I, with all reverence, try to attune our thoughts to the time conceptions required in this mighty theme by quoting those noble lines of the hymn—

"A thousand ages in Thy sight
Are like an evening gone,
Short as the watch that ends the night,
Before the rising sun."

Chapter 2

THE PROBLEM STATED

When we raise our eyes to the heavens on a clear night, thousands of bright objects claim our attention. We observe that all these objects move as if they were fastened to the inside of an invisible sphere. They are seen gradually ascending from the east, passing across the south, and in due course sinking towards the west. The sun and the moon, as well as all the other bodies, alike participate in this great diurnal movement. The whole scheme of celestial objects seems to turn around the two points in the heavens that we call the Poles, and so far as the pole in the northern hemisphere is concerned, its position is most conveniently indicated by the proximity of the well-known Pole Star.

Except this great diurnal motion, the vast majority of the bodies on the celestial sphere have no other movement easily recognisable, and certainly none which it is necessary for us to consider at present. The groups in which the stars have been arranged by the poetical imagination of the ancients exist to-day, as they have existed during all the ages since they were first recognised, without any noticeable alteration in their lineaments. The stately belt of Orion is seen to-night as Job beheld it thousands of years ago; the stars in the Pleiades have not altered their positions, relatively to the adjacent stars nor their arrangement among themselves, since the time when astronomers in early Greece observed them. All the bodies which form these groups are therefore known as fixed stars.

But besides the fixed stars, which exist in many thousands, and, of course, the sun and the moon, there are other celestial objects, so few in number as to be counted on the fingers of one hand, which are in no sense fixed stars. It is quite true that these wandering bodies, or planets, as they are generally designated, bear a certain resemblance to the fixed stars. In each case the star or the planet appears as a bright point, like many other bright points in the heavens, and star and planet both participate in the general diurnal motion. But a little attention will show that while the stars, properly so called, retain their relative places for months and years and centuries, the planets change their places so rapidly that in the course of a few nights it is quite easy to see, even without the aid of any instrument, that they have independent motion.

We may compare the movements of these bodies to the movement of the moon, which nightly shifts her place over a long track in the sky; and although we are not able to see the stars in the vicinity of the sun, inasmuch as the brilliant light of the orb quenches the feeble radiance from such stars, there is no doubt that, did we see them, the sun itself would seem to move relatively to the stars, just as does the moon and just as do the planets.

The fundamental distinction between stars and planets was noticed by acute observers of Nature in the very earliest times. The names of the planets come to us as survivals from the time when the sun, the moon, and the stars were objects of worship, and they come to us bearing the names of the deities of which these moving globes were regarded as the symbols. But it was not the movements of the planets alone which called for the notice of the early observers of the skies. The brightness and certain other features peculiar to them also attracted the attention of the primitive astronomers. They could not fail to observe that when the beautiful planet Venus was placed so as to be seen to the greatest advantage, her orb was far brighter than any other object in the host of heaven, the sun and the moon both of course excepted. It was also obvious that Jupiter at its best exceeded the stars in lustre, and sometimes approached even to that of

Venus itself. Though Mercury was generally so close to the sun as to be invisible among its beams, yet on the rare occasions when that planet was seen, just after sunset or just before sunrise, its lustre was such as to mark it out as one of the remarkable bodies in the heavens.

Thus the astronomers of the earliest ages pointed to the five planets and the sun and the moon as the seven wandering stars. The diligent attention of the learned of every subsequent period was given to the discovery of the character of their movements. The problems that these motions presented were, however, so difficult that not until after the lapse of thousands of years did their nature become understood. The supreme importance of the earth appeared so obvious to the early astronomers that it did not at first occur to them to assign to our earth a position which would reduce it to the same class as any of the celestial bodies. The obviously great size of our globe, the fact that to the uninstructed senses the earth seemed to be at rest, while the other bodies seemed to be in motion, and many other analogous circumstances, appeared to show that the earth must be a body totally different from the other objects distributed around us in space. It was only by slow degrees, and after much observation and reflection, and not a little controversy, that at last the true nature of our system was detected. Those who have been brought up from childhood in full knowledge of the rotation of the earth and of the other fundamental facts relating to the celestial sphere, will often find it difficult to realise the way such problems must have presented themselves to the observers of old, who believed, as for centuries men did believe, that the earth was a plane of indefinite extent fixed in space, and that the sun and the planets, the moon and the stars, were relatively small bodies whose movements must be accounted for as best they could be, consistently with the fixity and flatness of the earth.

But at last it began to be seen that the earth must be relegated to a position infinitely less important than that which the untutored imagination assigned to it. It was found that the earth was not an indefinite plane; it was rather a globe poised in space, without direct material support from any other body. It was found that the earth was turning round on its axis: while instead of the sun revolving around the earth, it was much more correct to say that the earth revolved around the sun.

Figure 5. Jupiter (May 30th, 1899, 10h. 9.5m.; g.m.t.). *(E. M. Antoniadi.)*

The astonishing truth was then disclosed that the five planets, Jupiter and Saturn, Mercury, Venus and Mars, stood in a remarkable relation to the earth. For as each of these planets was found to revolve round the sun, and as the earth also revolved round the sun, the assumed difference in character between the earth and the planets tended to vanish altogether. There was in fact no essential difference. If indeed the earth was smaller than Jupiter and Saturn, yet it was considerably greater and heavier than Mars or Mercury, and it was almost exactly the same size and weight as Venus. There was clearly nothing in the question of bulk to indicate any marked difference between our earth and the planets. It was also observed that there was no distinction to be drawn between the way in which the earth revolved round the sun and the movements of the plancts. No doubt the earth is not so near the sun as Mercury; it is not so near the sun as even Venus; on the other hand the sun is nearer the earth than Mars, while Jupiter is a long way further off than Mars, and Saturn is even beyond Jupiter again. It is these considerations which justify us in regarding our earth as one of the planets. We have also to note the overwhelming magnitude of the sun in comparison with any one of the planets. It will suffice to give a single illustration. The sun is more than a thousand times as massive as Jupiter, and Jupiter is the greatest of the planets. This latter noble globe is in fact greater than all the rest of the planets put together.

But before we can fully realise the circumstances of the solar system, it will be necessary to see how the stars, properly so called, enter into the scheme of things celestial. The stars look so like the planets that it has not infrequently happened that even an experienced astronomer has mistaken one for the other. The planet Mars is often very like the star Aldebaran, and there are not a few first-magnitude stars which on a superficial view closely resemble Saturn. But how great is the intrinsic difference between a star and a planet! In the first place we have to note that every planet is a dark object like this earth of ours, possessing no light of its own, and dependent entirely on the sun for the supply of light by which it is illumined. But a star is totally different. The star is not a dark object, but is really an object which is in itself intensely luminous and brilliant; the star is in fact a sun-like body. How then, it may well be asked, does a star like Aldebaran, which is indeed a sun-like body, and in all probability is quite as large and quite as brilliant as the sun itself, bear even a superficial resemblance to an object like Mars, which would not be visible at all were it not for the illumination with which the beams from the sun endow it?

The explanation of this striking resemblance is to be sought in the relative distances of the two objects. A light which is near to the eye may produce an effect quite as great as a very much stronger light which is further away. The intensity of a light varies inversely as the square of the distance. If the distance of a light from the eye be doubled, then the intensity of that light is reduced to one-fourth. Now Aldebaran as a sun-like body emits light which is literally millions of times as great as the gleam of sunshine which starts back to us after reflection from Mars; but Aldebaran is, let us say, a million times as far away from us as Mars, and this being so, the light from Aldebaran would come to us with only a million-millionth part of the intensity that it would have if the star were at the same distance as the planet. There can be no doubt that if Aldebaran were merely at the same distance from the earth as Mars, then Aldebaran would dispense lustre like a splendid sun. By moving Aldebaran further off its light, or rather the light that arrives at the earth, will gradually decrease until by the time that the star is a million times as far as Mars, the light that it sends us is about equal to that of Mars. If it were removed further still, the light that it would send us would become less than that which we receive from Mars, and if still more remote, Aldebaran might cease to be visible altogether.

This illustration will suffice to explain the fundamental difference between planets and stars, notwithstanding the fact that the two classes of bodies bear to each other a resemblance which is extremely remarkable, even if it must be described as being in a sense accidental. But we now know that all of the thousands of stars are to be regarded as brilliant suns, some of which may not be so far off as Aldebaran, though doubtless some are very much further. The actual distances are immaterial, for the essential point to notice is that the five planets are distinguished from the stars, not merely by the fact that they are moving, while the stars are at rest, but by the circumstance that the planets are comparatively close to each other and close to the sun, while the stars are at distances millions of times as great as the distances which the planets are from each other and from the sun.

We are now enabled to place the scheme of things celestial in its proper perspective. I shall suppose that at a point in a field in the centre of England, somewhere near Leamington, let us say, we drive in a peg to represent the sun. Let us draw a circle with that peg as centre, a yard being the radius, and let that circle represent the track in which the earth goes round the sun. I do not indeed say that the orbit of the earth is exactly a circle, and the actual shape of that orbit we may have to refer to later. As, however, the apparent size of the sun does not greatly alter with the seasons, it is evident that the track which our earth pursues cannot be very different from a circular path. Inside this circle which we have drawn with a yard radius, we shall put two smaller circles which are to represent the path in which Venus moves, and the path in which Mercury moves. Outside the path of the earth we shall draw another circle with a radius of five yards; this will be the highway along which the majestic Jupiter wends his way. Inside the path of Jupiter we shall put a circle which will represent the track of Mars, and outside the path of Jupiter a circle with ten yards as radius will represent the track of Saturn. In each of these circles we shall suppose the corresponding planet to revolve, and the time of revolution will of course be greater the further the planet is from the sun. To complete one of its circuits the earth will require a year, Jupiter twelve years, while Saturn, which in the ancient astronomy moved on the frontier of the solar system, will need thirty years to accomplish its mighty journey.

We have thus obtained a plan of the solar system; but now we should like to indicate the positions which some of the stars are to occupy on the same scale. Let us, to begin with, see where the very nearest fixed star is to be placed. We may suppose that the field at the centre of England, in which our little diagram has been constructed, is a large one, so that we can represent the places of objects which are ten or twenty times as far from the sun as Saturn. It is, however, certain that no actual field would be large enough to contain within its bounds the points which would faithfully represent the positions of even the nearest fixed stars. The whole county of Warwick would not be nearly big enough for this purpose; indeed we may say that the whole of England, or indeed of the United Kingdom, would not be sufficiently extensive. If we represented the star at its true relative distance, it could not be put down anywhere within the bounds of the United Kingdom; the nearest object of this kind would have to be far away out on the continent of Europe, or far away out on the Atlantic Ocean, far away down near the equator, or far away up near the pole. This illustration will at all events give some notion of the isolated position of the sun, with the planets revolving around it, in relation to the rest of the host of heaven.

We thus learn that the real scheme of the universe is widely different from that which a superficial glance at the heavens would lead us to expect. We are now able to put our system into its proper perspective. We are to think of the universe as consisting of a myriad suns, each sun, however, being so far from the other suns that viewed from any one of its neighbours it appears only of star-like insignificance. Let us fix our attention on one of these suns in space, and imagine that around it, and comparatively close to it, there are a number of small particles in revolution, the particles being illumined by the light and warmed by the heat of the central body to which they are attached. Viewed from one of those particles, the sun to which they belong would doubtless appear as a great and glorious orb, while a glance from one of these particles to any of the other myriad suns in space will show these orbs reduced to mere points of stellar light by reason of their enormous distance. This sun and the particles around it, by which of course we shall understand the planets, constitute what we know as the solar system. This illustration may suffice to show the isolation of our system in space, and that isolation is due to the vast distances by which the sun and its attendant worlds are separated from the myriads of other bodies which form the sidereal heavens. We must next, so far as our present subject requires it, consider the laws according to which the planets belonging to that system revolve around the sun.

Let us think first of a single one of these bodies which, as is most natural, we shall take to be the earth itself, and now let us consider by what agency the movement of the earth around the sun is guided along the path which so closely resembles a circle. It must, of course, be borne in mind that there can be no direct material connection between the two bodies; there is no physical bond uniting the earth to the sun. It is, however, certain that some influence proceeding from the sun does really control the motion. We may perhaps illustrate what takes place in the following manner. Here is a globe, and here in my hand I hold a tennis ball, which is attached to a silken thread, the other end of the thread being attached to the ceiling. The tennis ball is to hang so that both globe and ball are about the same height from the floor. We put the globe directly underneath the point on the ceiling from which the silken thread hangs. If I draw the tennis ball aside and simply release it, then of course everybody knows what happens—it is hardly necessary to try the experiment—the tennis ball falls at once towards the globe and strikes it. We may, if wc please, regard that tendency of the tennis ball towards the globe as a sort of attraction which the globe exercises upon the ball. I must, however, say that this is not a strictly accurate version of what actually takes place. The attraction of the earth for the tennis ball is of course largely neutralised by the support given by the silk thread. There is thus only a slight outstanding component of gravitation acting on the ball, and this component, which is virtually the effective force on the ball, tends to draw the ball directly towards the globe. For the purpose of our illustration we may neglect the direct attraction of the earth altogether; we may omit all thought of the tension of the silken thread. If there were indeed no attraction from the earth, the tennis ball might remain poised in space without falling; and if it were then attracted by the globe it would fly towards the globe just as we actually see it do. We are therefore justified in regarding the movement of the tennis ball as equivalent to that which would be produced if an attractive virtue resided in the globe by which it pulled the tennis ball. We may also imagine that the globe attracts the tennis ball in all its positions; for whatever be the point at which the ball is released it starts off straight towards the globe. This is our first experiment in which, having withdrawn the ball, it is merely released without receiving an initial impulse to one side.

Let us now try a different experiment. We withdraw the ball, and, instead of merely releasing it quietly and allowing it to drop directly to the globe, we give it a little throw sideways, perpendicular to the line joining it to the centre of the globe. If we start it with the proper speed, which a few trials will indicate, the ball can be made actually to move in a circle round the globe. If the initial speed be somewhat different, the path in which the tennis ball moves will not be a circle; it will rather be an ellipse of some form. Even if the speed be correct the orbit will always be an ellipse if the direction of the initial throw be not perpendicular to the line joining the ball to the centre of the globe.

Figure 6. Nebulous Region and Star-Cluster; (n.g.c. 2237-9 in Monoceros). (*Photographed by Dr. Isaac Roberts, F.R.S.*)

We can make the ball describe a very long ellipse or an ellipse which differs but little from a circle. But I would ask you to note particularly that, no matter how we may start the tennis ball into motion, it will, so long as it passes clear of the globe, move in an ellipse of some kind; but in making this statement we assume that a circle is a particular form of the ellipse.

And now for the lesson which we are to learn from this experiment, which, as it is so easily performed, I would wish everyone to try for himself. We have in this simple device an illustration of the movement of a planet around the sun. We see that this tennis ball can be made to move in a circle round the globe, and that as it performs this circular movement the globe is all the time attracting the ball towards it. Thus we illustrate the important law that when one body moves round another in a circular path this movement takes place in consequence of a force of attraction constantly exerted between the large body in the centre and the body revolving round it.

The principle here involved will provide the explanation of the movements of the planets round the sun. Each of the planets revolves round the sun in an orbit which is approximately circular, and each of the planets performs that movement because it is continually attracted by the sun. It is, however, necessary to add that there is a fundamental difference between the attraction of the sun for the planets and the attraction which the globe appeared to exert on the tennis ball in our experiment. The difference relates to the character of the forces in the two cases. If the tennis ball be drawn but a very small distance from the globe, the attraction between the two bodies is very slight. If the tennis ball be drawn to a greater distance from the globe, the attraction is increased correspondingly; and, indeed, in this experiment the attraction between the two bodies increases with the distance, and is said to be proportional to the distance.

But the case is very different in that particular kind of attraction by which the sun controls the movements of the planets. This attraction of gravitation, as it is called, also depends on the distance between the two bodies. But the attraction does not increase when the distance of the two bodies increases, for the change lies the other way. The attraction, in fact, diminishes more rapidly than the distance increases. If the distance between the sun and a planet be doubled, then the attraction between the two bodies is only a fourth of what the attraction was between the two bodies in the former case. This difference between the law of attraction as it exists in the solar system and the law of attraction which is exemplified in our little experiment produces a remarkable contrast in the resulting movements. The orbit in each case is, no doubt, an ellipse, but in the case of the tennis ball revolving round the globe the ellipse is so circumstanced that the fixed attracting body stood at its centre, while in the case of a planet revolving round the sun the conditions are not so simple. The sun does not stand in the centre of the ellipse. The sun is placed at that remarkable point of the ellipse so dear to the heart of the geometer, which he calls the focus.

The solar system consists, first, of the great regulating orb, the sun; then of the planets, each of which revolves in its own track round the sun; each of these tracks is an ellipse, and all these ellipses have this in common, that a focus in each is identical with the centre of the sun. In other respects the ellipses may be quite different. To begin with, they are not in the same plane, though it is most important to notice, as we shall have to discuss more fully hereafter, that these planes are not very much separated. The dimensions of the ellipses vary, of course, for the different planets, and the periods that the planets require for their several revolutions are also widely different in the cases of the different bodies; for the greater the diameter of a planet's orbit, the longer is the time required for that planet to complete a single journey round the sun. The sun presiding at the common focus of the orbits while governing the planets by its attraction, at the same time that it illumines them with its light and warms them by its rays, gives the conception of the solar system.

But the planetary system I have here indicated is merely that system as known to the ancients. It is very imperfect from the standpoint of our present knowledge. The solar system as we now know it, when telescopes have been applied with such marvellous diligence and success to the discovery of new bodies, is a system of much greater complexity. To the five old planets have been added two new and majestic planets—Uranus and Neptune—which revolve outside the track of Saturn. Hundreds of smaller planets, invisible to the unaided eye, the asteroids as they are called, also describe their ellipses round the presiding luminary. And then just as the sun controls the planets revolving round it, so do many of the planets themselves preside over subordinate systems of revolving globes. Our earth has a single attendant, the moon, which, under the guidance of the earth's attraction, performs its monthly journey; Jupiter has its fivc moons, while Mars has two, and Saturn eight or nine, besides his incomparable system of rings, and we must also add that Uranus has four satellites and Neptune one. To complete the tale of bodies in the solar system, we should add many thousands of comets, not to mention their more humble associates the meteors, which swarm in countless myriads. Finally, we are to remember that this elaborate system associated with the sun is an isolated object in the universe; it is but as a grain of sand in the extent of infinite space.

As we contemplate a system so wonderful, the question naturally arises, How came that system into being? We have to consider whether the laws of nature as we know them afford any rational explanation of the manner in which this system came into existence, any rational explanation of how the sun came to shine, how the earth had its beginning, how the planets came to revolve round the sun, and to rotate on their own axes. We have to seek for a rational explanation of the rings of Saturn, and of the satellites by which so many planets are attended. We have to show that a satisfactory explanation of these remarkable phenomena is forthcoming, and that it is provided by the famous doctrine of evolution, which it is the object of these lectures to discuss.

Chapter 3

The Fire-Mist

We commence this chapter with a scrutiny of the heavens, to see whether, among the bodies which it contains, we can discover any which appear at this moment to be in the condition through which our system has passed in some of its earlier stages.

So far as our unaided vision is concerned, we can see little or nothing in the skies which will render us assistance in our present endeavour. The objects that we do see in thousands are, of course, the stars, and, as we have already pointed out, the stars are sun-like objects, and as such have advanced many stages beyond the elementary condition. The stars are therefore not immediately available for the illustration we require. But when we come to look at the heavens through our telescopes we presently find that there are objects which were not visible to the eye, and which are neither stars nor planets. Closer examination of these objects with the powerful instruments of modern observatories, and especially with the help of those marvellous appliances which have enabled us to learn the actual chemistry of the heavenly bodies, supplies the suggestions that are required.

For not only does the telescope reveal myriads of stars which the naked eye cannot detect; not only does it reveal wonderful clusters in which thousands of stars are grouped closely together so as to form spectacles of indescribable magnificence, when we take into account the intrinsic splendour of each star-like point, but it also reveals totally different objects, known as nebulæ. These objects are not stars and are not composed of stars, but are vast extensions of matter existing in a far more elementary condition. It is to these curious bodies that we invite special attention at present. It is believed that they offer a remarkable illustration of the origin of the solar system. We shall first consider the best known object of this class. It is the Great Nebula in Orion.

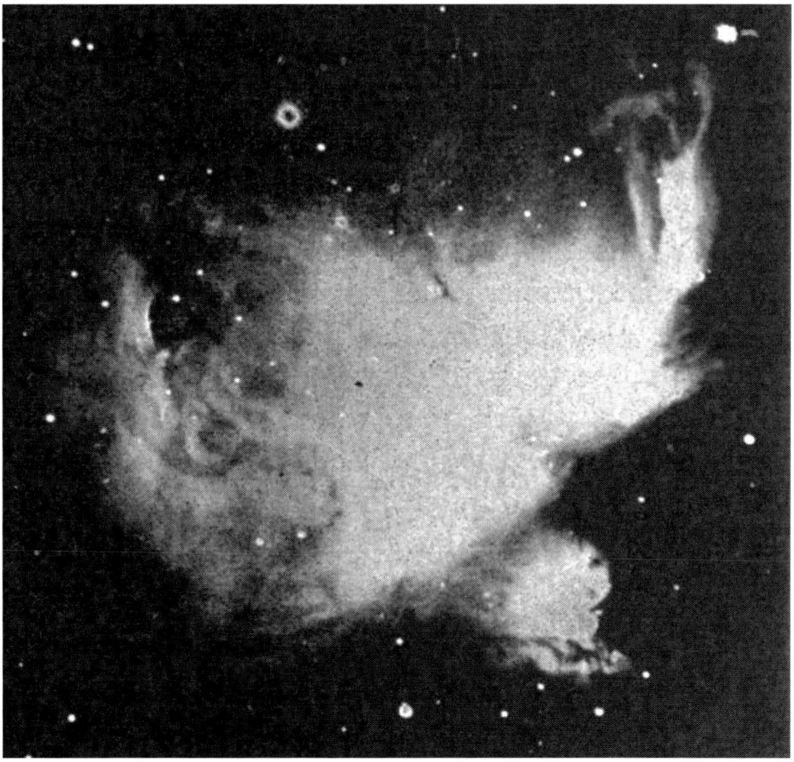

Figure 7. The Great Nebula in Orion (Lick Observatory, California).
(*From the Royal Astronomical Society Series.*)

And here it may be well to give an estimate which will enable us to form some notion of the size of this object. We are accustomed to recognise the stars as presenting the appearance of mere points of light; but an object like the Great Nebula stretches over a wide area of the sky. As to the actual extent of the space which it occupies we cannot speak with confidence. The fact is that with every increase in the power of the telescope the nebula appears to encroach more and more on the darkness of space around. We give in Figure 7 a representation of the Great Nebula as it appears on a photographic plate obtained at the Lick Observatory in California. But no picture can adequately represent the extraordinary delicacy of the object and the softness and tenderness with which the blue nebulous light fades into the black sky around. And it must not be imagined that the nebula, as seen on this picture, represents the utmost limits of the object itself. Every prolongation of the exposure, every increase in the sensitiveness of the plate, show more and more the extent of the nebula.

We shall, I doubt not, still be within the bounds of truth if we say that the nebula extends over an area ten times as great as that represented in this photograph. But we will take only the area of the object as shown in the photograph for the purpose of our calculation. Let us say that the nebula, as it is here represented, covers about two degrees square. I shall not attempt to express in miles the dimensions of an object so vast. I will try to give a conception of the size of the Great Nebula in a different manner. Let us employ the dimensions of our solar system for the purpose of comparison. Let us suppose that we draw, upon the scale of this celestial photograph, a map which shall represent the sun in the centre, the earth at her proper distance from the sun, and Jupiter in his orbit, which is five times the diameter of the earth's orbit; and then let us mark the other planets at their respective distances, even to Neptune, revolving in his great ellipse, with a diameter thirty times that of the earth's orbit. Let us then take the area of the orbit described by Neptune as a unit with which to measure the size of the Great Nebula in Orion. We shall certainly be well within the actual truth if we say that a million circles as big as that described by Neptune would not suffice to cover the area that is represented on this photograph. This will give some idea of the imposing dimensions of the Great Nebula in Orion.

But I would not have it to be supposed that the Great Nebula in Orion is unique, unless in respect to its convenient position. The circumstances of its situation in space happen to make it a comparatively easy object for observation by dwellers on the earth. There are, however, very many other nebulæ, although, with one exception—namely, the Great Nebula in Andromeda, to which we shall have to refer in a later chapter—they do not from our point of observation appear to be so brilliant as the nebula in Orion. The fact is that by large and powerful telescopes multitudes of these nebulæ are revealed, and the number ever tends to increase as greater depths in space are sounded. Many of the nebulæ are objects which possess sufficient detail to merit the particular attention which they receive from astronomers. It must, however, be confessed that by far the greater number of these objects are so dimly discerned that it is impossible to study their individual characteristics.

Among the nebulæ which possess sufficient individuality to merit study for our present purpose, I must mention the so-called Dumb-bell. This most interesting object can be seen in any good telescope. It requires, however, as indeed do all such objects, an instrument of the highest power to do it justice; in these modern days, however, the eye observation of nebulæ through great telescopes has been superseded by the employment of the photographic plate. I may take this opportunity of mentioning that a photograph really shows more details in the nebula than can be perceived even by the most experienced eye when applied to the most powerful telescope placed in the most favoured situation as to climate. Those lovers of nature who desire to observe celestial objects through a great telescope, and have not the opportunity of gratifying their wishes, may perhaps derive consolation from the fact that a good photograph actually represents the object much better than any eye can see it. More of the nebula is to be seen by looking at the photograph than has actually been directly observed by any astronomer.

We have chosen the Dumb-bell (Figure 8) and the Great Nebula in Orion as characteristic examples of this remarkable class of celestial objects; but there are many others to which I might refer, some of which we represent in these pages. The Crab Nebula (Figure 3) and others have been distinguished by special names; but I must forbear to dwell further on them, and rather hasten to give the results of recent observations which have enormously extended our knowledge of the nebulous bodies in the universe.

Let me first explain the source whence this extraordinary accession to our knowledge has arisen. We owe it to the astronomers at the Lick Observatory, that remarkable institution placed on the summit of Mount Hamilton in California. Many important discoveries had already been made with the noble instruments with which the famous Lick Observatory had originally been endowed by its founder; it is, however, by a recent addition to its magnificent apparatus that the discoveries have been made which are specially significant for our present purpose.

Many years ago Dr. A. A. Common, the distinguished English astronomer, constructed an exquisite reflecting telescope of three feet aperture (Figure 9). With this telescope Dr. Common himself obtained notable results in photographing the heavens, and his success earned the award of the Gold Medal of the Royal Astronomical Society. This telescope passed into the possession of Mr. E. Crossley, of Halifax, and sometime later Mr. Crossley presented it to the Lick Observatory. The great mirror, after its voyage across the Atlantic, was duly erected on the top of Mount Hamilton, and fortunately for science Professor Keeler, whose early death astronomers of both continents greatly deplore, devoted himself to the study of the heavens with its aid. He encountered many difficulties, as might perhaps be expected in such a task as he proposed. His patience and skill, however, overcame them, and though death terminated his labours when his great programme had but little more than commenced, the work he had already accomplished has led to results of the most striking character. Of the skill that he obtained in photographing celestial nebulæ we have given illustrations in Figures 6 and 7.

It is not to the individual portraits of notable nebulæ that we are now about to refer. The most striking characteristic of the sidereal heavens is not to be found in the fact that in one part of the sky we have a brilliant Sirius, in another a Capella, and in a third a Canopus, but in the fact that the heavens wherever we may test them are strewn with incalculable myriads of stars, many of which appear faint only on account of their distance and not because they are intrinsically small. In like manner the remarkable fact with regard to the nebulæ which has been disclosed by Keeler's memorable researches with the Crossley Reflector is the existence not alone of the great nebulæ, but of unexpected scores of thousands of small nebulæ, or rather, I should say, of nebulæ which appear small, though doubtless in many cases these objects are intrinsically quite as splendid as the Dumb-bell Nebula or the Nebula in Orion. They only seem small in consequence of being many times further from us than are the more famous objects.

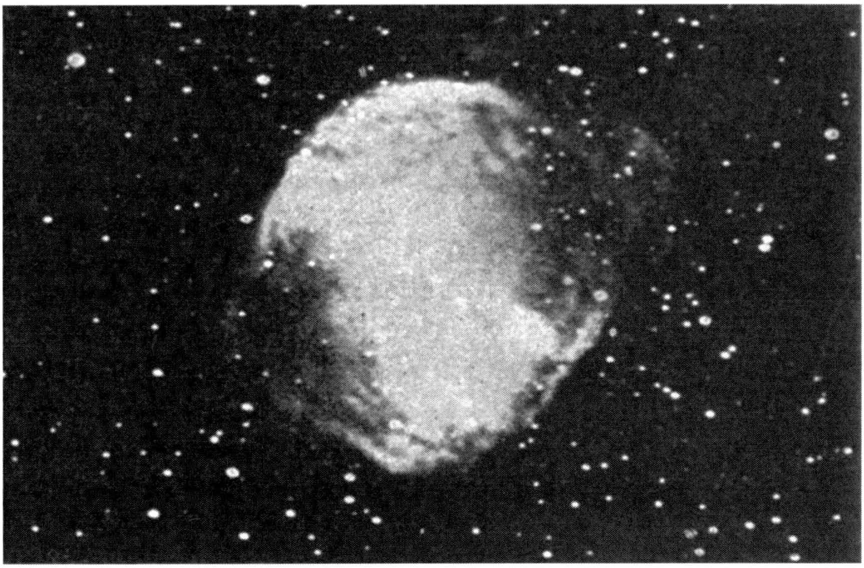

Figure 8.—The Dumb-bell Nebula (Lick Observatory, California). (*From the Royal Astronomical Society Series.*)

Professor Keeler's experience was a remarkable one. He was photographing a well-known nebula with the Crossley Reflector, and he was a little surprised to find that on the same plate which gave him the nebula at which he was aiming there were no fewer than seven other small nebulous objects previously unknown to astronomers. It at first appeared to him that this must be an unusual number of nebulæ to find crowded together on one plate which covered no more than one square degree of the heavens, an area about five or six times as large as the area of the full moon. Subsequent experience, however, showed him that this fact, however astonishing, was not at all unusual. In fact, he found to his amazement that, expose the plate where he pleased, he generally obtained new nebulæ upon it, and sometimes even a much larger number than the seven which so greatly surprised him at first. I may mention just one or two instances. There is a well-known and interesting nebula in Pegasus which Professor Keeler photographed. When he developed the plate, which, of course, included a considerable region of the heavens in the vicinity of the particular nebula, he found to his astonishment that, besides the nebula he wanted, there were not less than twenty other nebulæ on the plate. But there is a more striking instance even than this. A plate directed to a part of the constellation of Andromeda, with the object of taking a portrait of a particular nebula of considerable interest, was found to contain not only the desired nebula, but no fewer than thirty-one other new nebulæ and nebulous stars. Nor have we in these statements exhausted the nebulous contents of these wonderful plates, if indeed we have rightly interpreted their nature. Professor Keeler tells us that he finds upon them a considerable number of objects which in all probability are also nebulæ, though they are so small that the telescope is unable to reveal them in their true character. Examination does little more than show these objects as points of light which, however, are apparently not stars.

In the remarkable paper from which I have taken these facts Professor Keeler makes an estimate which is founded on the examination of his plates. If the heavens were to be divided into panels, each one square degree in area, there would be about forty thousand panels. It follows that if we desired to photograph the whole heavens, and if each of the plates was to cover one square degree, forty thousand pictures would be needed for the representation of the whole celestial sphere. Keeler's work convinced him that such plates taken by the Crossley Reflector would, on an average, each show at least three new nebulæ. He admitted it is quite possible that there may be regions of the sky in which no new nebulæ are to be found.

Figure 9. The Crossley Reflector (Constructed by Dr. A. A. Common F.R.S. and now at the Lick Observatory).

But in the regions which he had so far tested he invariably found more than three nebulæ on each square degree; indeed, as we have seen, on some of his plates he found a much larger number of these remarkable objects. He therefore said that he makes but a very moderate estimate when he gives a hundred and twenty thousand as the probable number of the new nebulæ within the reach of the photographic plates of the Crossley Reflector. The enormous extension which these investigations have given to our knowledge demands the serious attention of all interested in the heavens. The discoveries of the earlier astronomers had led to the knowledge of about six thousand nebulæ; the Crossley Reflector at the Lick Observatory has now rendered it practically certain that the number of nebulæ in the heavens must be at least twenty-fold as great as had been hitherto supposed.

In subsequent chapters we are to present the evidence for the belief that this earth of ours, as well as the sun and all the other bodies which form the solar system, did once originate in a nebula. According to this view the materials which at present are found in the globes of the solar system were once distributed over a vast extent of space as a fire-mist, or nebula. It is surely very pertinent to be able to show that a nebula, such as we suppose to have been the origin of our system, is not a mere figment of the imagination. No doubt it is impossible for us now to show the original nebula from which the solar system has been evolved. It is nevertheless possible, as we have seen, to show that a hundred and twenty thousand nebulæ are now actually existing of every grade of magnitude. They range from such magnificent objects as the Great Nebula in Orion and the Dumb-bell Nebula, down to objects wholly invisible, not merely to the unaided eye, but even in the most powerful telescope, and only to be discerned as hazy spots of light on the photographic plates of an instrument such as the Crossley Reflector.

Though no eye has seen the actual stages in the grand evolution of our solar system, we may at least witness parallel stages in the evolution through which some of the myriads of other nebulæ are now passing. We find some of these nebulæ in that excessively diffused condition in which they are devoid of visible structure. Material in this form may be regarded as the primæval nebula. There is at least one of these extraordinary objects which is larger a great deal than even the Great Nebula in Orion, but altogether too faint to be seen except by the photographic plate. Here we find, as it were, the mother-substance in its most elementary stage of widest possible diffusion, from which worlds and systems, it may be, are yet to be evolved. From diffused objects such as shown in Figure 9 we can pass to other nebulæ in which we see a certain advance being made in the process by which the nebula is transformed from the primitive condition. We can point to yet other nebulæ in which the advance to a further stage of development is more and more pronounced. Thus the various stages in the evolution of a system are to be witnessed, not indeed in the transformation of a single nebula, but by observing a properly arranged series of nebulæ in all gradations, from the diffused luminous haze to a star with a faint nebulous surrounding. Such was Herschel's original argument, and its cogency has steadily increased from the time he first stated it down to the present hour.

Chapter 4

NEBULÆ—APPARENT AND REAL

There is perhaps hardly any telescopic object more pleasing or more instructive than a globular cluster of stars when viewed through an instrument sufficiently powerful to do justice to the spectacle. There are several star-clusters of the class designated as "globular." The most famous of these, or, at all events, the one best known to northern astronomers, is found in the constellation of Hercules, and is for most purposes sufficiently described by the expression, "The Cluster in Hercules." The genuine lover of Nature finds it hard to withhold an exclamation of wonder and admiration when for the first time, or even for the hundredth time, the Cluster in Hercules is adequately displayed in the field of a first-class telescope.

In Figure 10 is a photograph of this celebrated object, which was taken by Dr. W. E. Wilson, F.R.S., at his observatory at Daramona, in Ireland. The picture has been obtained from an enlargement of the original photograph taken with the telescope in Mr. Wilson's observatory. It is, however, precisely as Nature has given it, except for this enlargement. You will note that towards the margin of the cluster the several stars are seen separately, and in many cases with admirable distinctness. We do, however, occasionally find two or more stars so close together that their images overlap; and, indeed, in the centre of the cluster the stars are so close together that it is impossible to differentiate them, so as to see them as individual points of light. We need have no doubt, however, that the cluster is mainly composed of separate stars, although the difficulties interposed by our atmosphere, added to the necessary imperfections of our appliances, make it impossible for us to discriminate the individual stars.

In looking at a star group of this particular kind the observer may perhaps be reminded of a swarm of bees in flight from the hive, for the stars in the cluster are, on a vast scale, apparently associated in the same way as the bees, on a small scale, are associated in the swarm. We may also compare the stars in the cluster to the bees in the swarm in another respect. Each bee in the swarm is in incessant movement. There can be no doubt that each star in a globular cluster is unceasingly changing its position with reference to the others. The distance by which the cluster is separated from the earth renders it impossible for us to see those movements, at all events within those narrow limits of time over which our observations have as yet extended; but the laws of mechanics assure us that the mutual attraction of the stars in this cluster must give rise to incessant movements, and that this must be the case notwithstanding the fact that the relative places of the stars in the cluster show no alteration that can be recognised from one year's end to another.

I may, however, mention that though there may be no movements in these stars great enough to be observed, yet the brightness of some of them shows most remarkable fluctuations. The investigations of Professor Bailey and other astronomers have, indeed, disclosed such curious variability in the brightness of some of these stars that if it were not for the exceedingly high authority by which this phenomenon has been guaranteed we should, perhaps, almost hesitate to believe so startling a fact. It has, however, been most certainly proved that many of the stars in certain globular clusters pass through a series of periodical changes of lustre. The period is a very short one as compared with the periods of better known variable stars, for in this case twenty-four hours are more than sufficient for a complete cycle of changes, and it not infrequently happens that in the course of a single quarter of an hour a star will lose or gain brightness to the extent of a whole magnitude. The phenomenon referred to is at the present moment engaging the careful attention of astronomers; but it offers a problem of which, indeed, it is not at present easy to see the solution.

Figure 10. The Cluster in Hercules.
(Photographed by Dr. W. E. Wilson, F.R.S.)

Our immediate concern, however, with the globular star-clusters relates to a point hardly of such refinement as that to which I have just referred; it is one of a much more elementary nature. The photograph in the figure may be considered to represent the Cluster in Hercules as it would be seen with a telescope of very considerable visual power, for the object would assume a different appearance in a telescope which was not first class. The perfection of a really powerful instrument is tested by its capability of exhibiting as two separate points a pair of stars which are excessively close together, and which in an instrument of inferior power cannot be distinguished, but seem fused into a single object. The defining power of a telescope—that is to say, its capability for separating close double stars—is increased with the size of the instrument, always granting, of course, that there is equal optical perfection in both cases. It follows that the more powerful the telescope the more numerous are the stars which can be seen separately in a globular cluster.

If, however, a small telescope be used, or a telescope which, though of considerable size, has not the high optical perfection that is demanded in the best modern instruments, then adjacent stars are not always to be seen separately. It may be that the telescope, on account of its small size, cannot separate the objects sufficiently, or it may be that the imperfections of the telescope do not present the star as a point of light, but rather as a more or less diffused, luminous disc. In either case it may happen that a star overlaps other stars in its immediate neighbourhood, and consequently an object which is really a cluster of separate stars may fail altogether to present the appearance of a cluster.

I have been alluding to something which, as every astronomer knows, is of practical importance in the observatory. Like everyone else who has ever used a telescope, I have myself seen the Cluster of Hercules with just the same misty appearance in a small telescope that an undoubted nebula possesses in the very finest instrument. It is, accordingly, sometimes impossible, merely by observation with a small instrument, to distinguish between what is certainly a cluster of stars and what is certainly a nebula. It has indeed not infrequently happened that an observer with a small telescope has discovered what appeared to him to be a nebula, and he has recorded it as such; and yet when the same object was subsequently examined with an instrument of greater defining power the nebulous character has been seen to have been wrongly attributed. The object in such a case is proved to be nothing more than a cluster of stars, of which the individual members are either intrinsically faint or exceedingly remote;

it certainly is not a mass of that fire-mist or gaseous material which alone is entitled to be called a nebula.

It is therefore a question of importance in practical astronomy to decide whether objects which appear to be nebulæ are really entitled to the name, or whether the nebulous appearance may not be an optical illusion. The operation by which an object previously deemed to be a nebula is shown by the application of increased telescopic power to be a cluster of stars is commonly known as the resolution of a nebula. About fifty years ago the mighty six-foot reflecting telescope of Lord Rosse, and other great instruments, were largely employed on this work. It was, indeed, at that time held to be one of the special tasks which came most legitimately within the province of the big telescopes, to show that the so-called nebulæ of earlier observers were resolvable into star-clusters under the superior powers now brought to bear upon them.

The success with which this process was applied to many reputed nebulæ, which were thereby shown to be not entitled to the name, led not unnaturally to a certain conjecture. It was admitted that certain objects which had successfully resisted the resolving powers of inferior instruments were forced to confess themselves as mere star-clusters when greatly increased telescopic power was brought to bear on them; and it was conjectured that similar success would attend the attempts to resolve still other nebulæ. It was even supposed that every object described as a nebula could only be entitled to bear that designation provisionally, only indeed until some telescope of sufficient power should have been brought to bear on it. It seemed not unreasonable to surmise that every one of the so-called nebulæ is a cluster of stars, even though a telescope sufficiently powerful to effect its resolution might never be actually forthcoming.

I do not, indeed, believe that this opinion as to the ultimate resolvability of all nebulæ could have been shared by those who had much practical experience in the actual observation of these objects with the great telescopes, for the particular classes of nebulæ which in telescopes of superior powers resolved themselves into groups of stars had a characteristic appearance. After a little experience the observer soon learned to recognise those nebulæ which promised to be resolvable. The object might not indeed be resolvable with the powers at his disposal, but yet from its appearance he often felt that the nebula would be probably resolved if ever the time should come that greater powers were applied to the task.

It is easy to illustrate the question at issue by the help of the photograph of the Cluster in Hercules in Figure 10. Each of the stars is there distinct, except where they are much crowded in the centre. If, however, the photograph be examined through one of those large lenses which are often used for the purpose, and if the lens be held very much out of focus, the stars will not be distinguishable separately, and the whole object will be merely a haze of light. This illustration may help to explain how the different optical conditions under which an object is looked at may exhibit, at one time as a diffused nebula, an object which in better circumstances is seen to be a star-cluster.

The astronomer who was fortunate enough to have the use of a really great telescope would not fail to notice that, in addition to the so-called nebulæ already referred to, which were presumably resolvable, there were certain other objects, generally characterised by a bluish hue, which in no circumstances whatever presented the appearance of being composed of separate stars. We now know for certain that these bluish objects are not clusters of stars, but that they are in the strictest sense entitled to the name of nebulæ, and that they are gaseous masses or mists of fire-cloud. The full demonstration of this important point was not effected until 1864.

The fact that so very many of the nebulæ were resolved led not unreasonably to the presumption that all the nebulæ would in due time also yield. But there were many who could not accept this view, and there was a long discussion on the subject. At last, however, the improvements in astronomical methods have cleared up the question. Sir W. Huggins has shown that there are two totally distinct classes of nebulæ, or rather of so-called nebulæ. There are certain nebulæ which can be resolved, and there are certain nebulæ which cannot. A nebula which can be resolved would be a veritable cluster of stars, and is not really entitled to the name of nebula; a nebula which cannot be resolved would be entitled to the name, for it is a volume of gas or of gaseous material which is itself incandescent. We have been provided with a beautiful criterion by which we can decide to which of these classes any nebulous-looking object belongs.

The spectroscope is the instrument which discriminates the two different classes of objects. This remarkable apparatus, to which we owe so much in every department of astronomy, receives the beam of light from the celestial body. The instrument then analyses the light into its component rays, and conducts each one of those rays separately to a distinct place on the photographic plate. When the photograph is developed we find on the various parts of the plate the evidence as to the class of rays which have entered into the composition of the light that has been submitted to this very searching form of examination.

The light which comes from a star or any star-like body, including the sun itself, may first be described. That light, after passing through the spectroscope and having been conducted to the photographic plate, will produce a picture of dark lines on a bright background; this is, at least, the spectrum which a star generally presents. There are, indeed, many types of stellar spectra, for there are many different kinds of stars, and each kind of star is conveniently characterised by the particular spectrum that it yields. If the star be one of small magnitude, then the lines in its spectrum may be detected, but only with great difficulty. It not infrequently happens that the photograph of the spectrum of such a star will show no more than a continuous band of light without recognisable lines; and this is what occurs in the case of a resolvable nebula, where the stars are so closely associated that the spectrum of each separate star cannot be distinguished. The spectrum of a resolvable nebula is merely a streak of light, which is the joint effect of all the spectra. The spectrum is then too faint to show the rainbow hues which present such beautiful features in the spectrum of a bright star, as they do in the spectrum of the sun itself.

I give, in the adjoining figure, portions of the photographs of two spectra of celestial objects. They have been taken from the Atlas of representative stellar spectra in which Sir William and Lady Huggins have recorded the results of their great labours. Two spectra are represented in this picture, the uppermost being the spectrum of the sun, while the lower and broader one is the spectrum of the bright star Capella. It has not been possible within the limits of this picture to include the whole length of these two spectra, and it must therefore be understood that the photographs given in the Atlas are each about five times as long as the parts which are here reproduced.

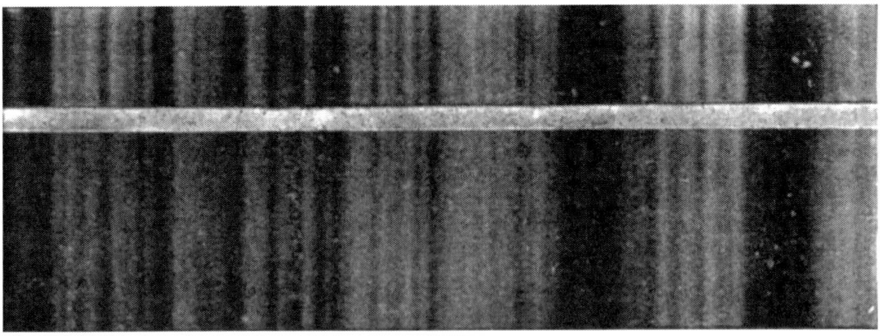

Figure 11. Sun and Capella. Sun above. Capella below.
(*Sir William and Lady Huggins.*)

But the characteristic portions of the spectra selected are sufficient for our present argument. It will be noted, first of all, that there is a singular resemblance between the details of the spectrum of the sun and those of the spectrum of the star. No doubt the breadth of the stellar picture in the lower line is greater than that of the solar picture in the upper line; but this point is not significant. The breadth of the spectrum of the sun could easily have been made as wide or wider if necessary. The breadth is immaterial, for the character of a spectrum is determined not by its breadth, but by those lines which cross it transversely. It will be seen that there are here a multitude of lines, some being very dark, and some so faint as to be hardly visible. Both spectra exhibit every variety of line, between the delicate marks which can barely be seen and the two bold columns on the right-hand side of the picture.

The characteristic of the spectrum is given by the number, the arrangement, the breadth, the darkness, and the definiteness of the lines by which it is crossed, and the first point that we note is the remarkable resemblance in these different respects between the two spectra. The lines are practically identical, at least so far as those parts of the spectrum represented in this picture are concerned. We have thus a striking illustration of the important fact, to which we have so often to make allusion, of the general resemblance of the sun to the stars. Not only do we know that if the sun were removed about a million times as far as it is at present its light would be reduced to that of a star, but that the star Capella transmits to us light consisting essentially of the same waves as those which enter into a beam of sunlight. No more striking illustration of the analogy between the sun and a star can be found than that which is given in this photograph from the famous Observatory at Tulse Hill.

But it must not be inferred that because the spectra of sun and star are like each other, they are therefore absolutely identical. There are many lines and details to be seen on the actual photographic plate which are too delicate to be reproduced in such copies as it is possible to make. When a close comparison is made on the actual plate itself of the lines in the solar spectrum and the lines in the spectrum of Capella, it is observed that, though they are the same so far as the more important lines are concerned, yet that there are many lines found in the spectrum of Capella which are not found in the spectrum of the sun.

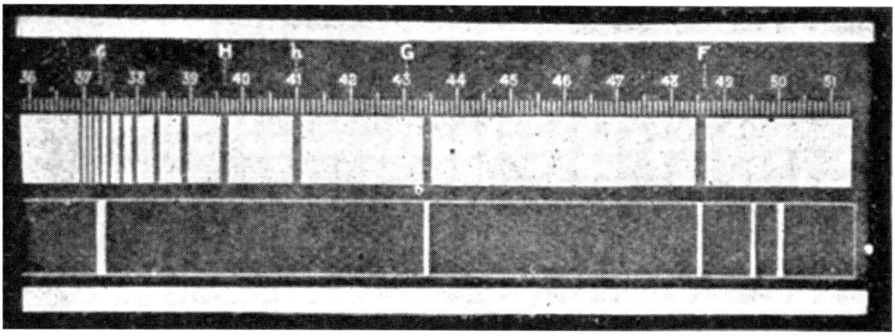

Figure 12. Spectrum of Nebula in Orion and Spectrum of a White Star.
(*Sir William Huggins, K.C.B.*)

The contrast between the spectrum of a nebula properly so called and the spectrum of a star is well illustrated by the accompanying picture, in which Sir W. Huggins exhibits the photograph of the spectrum of the Nebula in Orion in comparison with the spectrum of a star. The uppermost of the two is the spectrum of the star. It will be noted that this spectrum is very different from that which we have already seen in Capella. Instead of a vast multitude of lines resembling the lines of the solar spectrum, the spectrum of a star of the type here represented, of which we may take Sirius as the most striking example, exhibits but a few lines. We regard them as one system of lines, for we know they are physically connected. They are all alike due to the presence of a single element in the star, that element being in fact hydrogen. But though the spectra of Capella and Sirius are so totally different, the differences relate only to the distribution of the lines, and to their number, darkness, and width. In both cases we observe the characteristic of the light from an ordinary bright star, namely, that the spectrum is composed of a bright band with dark lines across it. It ought, perhaps, to be mentioned here that there are certain very special stars which do exhibit some bright lines in addition to a more ordinary spectrum; this is especially the case in the new stars which occasionally appear. Thus in the case of the new star which appeared in Perseus, in 1901, there were several remarkable bright lines. This most interesting object will be referred to again in a later chapter.

Widely different from the spectrum of any star whatever is the lower of the two spectra which are shown in the figure. This lower spectrum is that of the Great Nebula in Orion. At once we see the fundamental characteristic of a nebula; its spectrum exhibits five bright lines on a dark field. I do not say that the Great Nebula in Orion has not more than five lines; there are indeed many others, for Sir William Huggins has himself pointed out a considerable number, and the labours of other observers have added still more; but the five lines here set down are the principal lines. They are those most easily seen; the others are generally extremely delicate objects arranged in groups of five or six. But the lines which this picture shows are quite sufficient to exhibit that fundamental characteristic of the nebular spectrum, namely, a system of bright lines on a dark field. I may further mention that certain lines in the spectrum indicate the presence of the element hydrogen in the Great Nebula in Orion, and we owe to Dr. Copeland the interesting discovery that the remarkable element helium is also proved to exist in the nebula.

The pictures, at which we have been looking, will suffice to make clear the criterion, which astronomers now possess, for deciding whether an object which looks nebulous is really a gaseous nebula, or ought rather to be regarded as a star-cluster. If the object be a star-cluster, then the spectrum that it gives will be the resultant of the spectra of the stars, and this will be a continuous band of light. If the stars are bright enough, it may be that dark lines can be detected crossing the spectra, but in the case of the clusters it will be more usual to find the continuous band of light so faint that the dark lines, even if they are there, are not distinguishable.

If, on the other hand, the object at which we are looking, not being a cluster of stars, is indeed a mass of glowing gas, or true nebula, then the spectrum that it sends us is not the continuous spectrum such as we expect from the stars. The spectrum which the nebula proper transmits to the plate is said to be discontinuous. In some cases it is characterised by only a single bright line, and in others there may be two, or three, or four bright lines, or, as in the case shown in Figure 12, the number of bright lines may be as many as five. It may indeed happen, in the case of some exquisite photographs, that the number of lines in the spectrum of the nebula will be increased to a score or possibly more. There may also be faint traces of a continuous spectrum present, this being due to the stars scattered through the object, from which perhaps even the most gaseous nebula is not entirely free. But the characteristic type of nebular spectrum is that in which the bright lines, be they one, or few, or many, are separated by intervals of perfect darkness. When it is found that the spectrum of a nebula can be thus described, it is correct to say that the nebula is truly a gaseous object.

In the lists given by Scheiner in his interesting book, “Astronomical Photography,” the number of gaseous nebulæ is set down as seventy-three. Of course no one pretends that this enumeration is exhaustive. It claims to be no more than a statement of the number of nebulæ which have been proved, by observations made up to the present, to be of a gaseous description. Seeing that there are, as we have already stated, many scores of thousands of nebulous-looking objects, it is probable that the number above given is not more than a small fraction of the number of gaseous nebulæ actually within reach of our instruments.

It may, however, be assumed that more than half the objects which are called nebulæ are not of the gaseous type. This is a point of some importance, which appears to follow from the facts stated by Professor Keeler in connection with his memorable researches with the Crossley Reflector. In a later chapter we discuss important questions connected with what are called *spiral* nebulæ. We may, however, here record that no spiral nebulæ have as yet been pronounced gaseous. Professor Keeler assures us that, of the one hundred and twenty thousand nebulæ which he estimates to be within reach of the Crossley Reflector, far more than half are of the spiral character. If, then, we assume that the spectra of spiral nebulæ are always continuous, it seems to follow that less than half the nebulous contents of the heavens possesses the discontinuous spectrum which is characteristic of a gaseous object.

We are not entitled to assume that a nebula, or reputed nebula, which shows a continuous spectrum, must necessarily be a cluster, not merely of star-like bodies, but of bodies with masses comparable with those of the ordinary stars. Our argument does most certainly suggest that the body which yields a continuous spectrum is not a gaseous body; but it may be going too far to assert that therefore it is a cluster of stars in the ordinary sense. We do often find true nebulæ and star-clusters in close association. The Nebula in the Pleiades is an example.

It may be desirable to add a few words here as to the physical difference between a continuous spectrum and a discontinuous spectrum. The light from a body, known to be gaseous, shows through the prism the discontinuous spectrum of bright lines upon a dark background. If, on the other hand, a solid be raised to incandescence, such, for instance, as a platinum wire heated white-hot by an electric current, or a cylinder of lime submitted to an oxyhydrogen blowpipe, then the spectrum that it yields is continuous. All the colours of the rainbow, red, orange, yellow, green, blue, indigo, violet, are shown in such a spectrum as a continuous band of light, though the band is not crossed by dark lines. It would therefore appear that the continuous spectrum is characteristic of an incandescent solid, and the discontinuous spectrum of a glowing gas. But here it may be urged that the sun presents a difficulty. We so often refer to the spectrum of the sun as continuous, that it might at first appear as if the spectrum of the sun resembled that produced by radiation from a solid body. But, as is well known, the sun is not a solid body. Even if the sun be solid at the centre, it is certainly far from being solid in those superficial regions called the photosphere, from which alone its copious radiation is emitted. If the sun is not a solid body, how comes it to emit a radiation characterised in the same way as the radiation from a white-hot solid? Why does the solar spectrum not exhibit features characteristic of radiation from an incandescent gas? The point is well worthy of attention; it finds an explanation in the nature of the photosphere from which the sun's radiation proceeds.

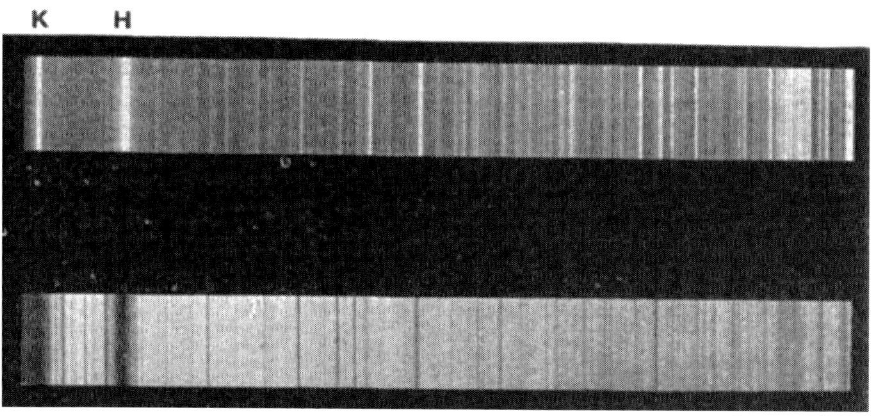

Figure 13. Solar Spectra with Bright Lines and Dark Lines during Eclipse. (*Photographed by Captain Hills, R.E.*)

The photosphere, though not, of course, to be described as a solid body, does not most certainly, so far as its radiation is concerned, behave like a gaseous body. In the glowing clouds of the photosphere the carbon, of which they are composed, is not in the gaseous form; it has passed into solid particles, and it is these particles, in the highest condition of incandescence, which emit the solar radiation. Although these particles are sustained by the gases of the sun, and are associated in aggregations which form the dazzling clouds of the photosphere, yet each one of them, in so far as its individual radiation is concerned, ought to be regarded as a solid body. The radiation from the sun is, therefore, essentially not the radiation from an incandescent gas; it is the radiation from a glowing solid. This is the reason why the solar spectrum is of the continuous type.

By the kindness of Captain Hills, R.E., I am able to show a photograph containing two spectra taken during a recent eclipse, which will serve as an excellent illustration of the different points which we have been discussing. It is, indeed, true that neither of the spectra, here referred to, belongs to nebulæ, whether genuine gaseous objects or not. Both of the spectra in Captain Hills' picture are actually taken from the sun. The conditions under which these spectra were obtained make them, however, serve as excellent illustrations of the different types of spectra. We are to notice that the upper band, which contains what is called the "flash" spectrum, exhibits bright lines on a dark background. See, for instance, the two lines so very distinctly marked, which are indicated by the letters H and K. These lines are very characteristic of the solar spectrum, and it may be mentioned that they are indications of the presence of a well-known element. These lines prove that the sun contains calcium, the metal of which common lime is the oxide. It is, indeed, the presence of this substance in the sun which gives rise to these lines. We shall refer again to this subject in a later chapter.

As the upper of the two spectra exhibit H and K as white lines on a dark background, so the lower represents the same lines as dark objects on a white background. These photographs give illustrations of spectra of the two different classes which provide means of discriminating between a genuine nebula and an object which, though it looks like a nebula, is not itself gaseous.

But, it will be asked, how can the spectra of the two distinct types both be obtained from the sun? The explanation of this point is an interesting one. The lower of the two is the ordinary solar spectrum; it is a continuous spectrum showing dark lines on a bright field. The upper spectrum, which shows bright lines on a dark field, is produced by a small part of the sun just at the moment when the eclipse is total. The circumstances in which that picture was secured will explain its character.

Figure 14. The Nebula in the Pleiades (Exposure 10 hours).
(*Photographed by Dr. Isaac Roberts, F.R.S.*)

The moon had completely covered that dazzling part of the sun which we ordinarily see, but a region of intensely glowing gaseous material in the sun's atmosphere was too high above the surface to be completely hidden by the moon. The spectrum of this region, consisting of the characteristic bright gaseous lines, is here represented. The ordinary light of the sun being cut off, opportunity was thus afforded for the production of the spectrum of the light from the glowing gas, and we see this spectrum to be of the nebular type.

And now we may bring this chapter to a close by calling attention to the very important bearing which its facts have on the Nebular Theory. It is essential for us to see how far modern investigation and discovery have tended either to substantiate or refute that famous doctrine which traces the development of the solar system from a nebula. To do this it is necessary to contrast the knowledge of nebulæ, as it exists at present, with the knowledge of nebulæ as it existed in the days of Kant and Laplace and Herschel.

We assuredly do no injustice to Kant or to Laplace if we say that their actual knowledge of the nebulous contents of the heavens was vastly inferior to that possessed by Herschel. There is not a single astronomical observation of nebulæ recorded by either Kant or Laplace; it may be doubted whether either of them ever even saw a nebula. Their splendid contributions to science were made in directions far removed from those of the practical observer, who passes long hours of darkness in the scrutiny of the celestial bodies. Herschel, on the other hand, was pre-eminently an observer. His nights were spent in the most diligent practical observation of the heavens, and at all times the nebulæ were the objects which received the largest measure of his attention, with the result that the knowledge of nebulæ received the most extraordinary development from his labours. Earlier astronomers had no doubt observed nebulæ occasionally, but with their imperfect appliances only the brighter of these objects were discernible by them. The astonishing advance made by the observations of Herschel is only paralleled by the advance made a hundred years later by the photographs of Keeler.

But it must be remembered that though Herschel observed nebulæ, and discovered nebulæ, and discoursed on nebulæ in papers which to this day are classics in this important subject, yet not to the last day of his life could he have felt sure that he had ever seen a genuine nebula. He might have surmised, and he did surmise, that many of the objects he set down as nebulæ were actually gaseous objects, but he knew that many apparent nebulæ were in truth clusters of stars, and he had no means of knowing whether all so-called nebulæ might not belong to the same category.

It was not till nearly half a century after Sir William Herschel's unrivalled career had closed that the spectroscope was invoked to decide finally on the nature of these mysterious objects. That decision, which has been of such transcendent importance in the study of the heavens, was not pronounced till 1864. In that year Sir William Huggins established the fundamental truth that the so-called nebulæ are not all star-clusters, but that the universe does contain objects which are most certainly gigantic volumes of incandescent gases.

This great achievement provided a complete answer to those who urged an objection, which seemed once very weighty, against the Nebular Theory. It must be admitted that before 1864 no one could have affirmed with confidence that any genuine nebula really existed. It was, therefore, impossible for the authors of the Nebular Theory to point to any object in the heavens which might have illustrated the great principles involved in the theory. The Nebular Theory required that in the beginning there should have been a gaseous nebula from which the solar system has been evolved. But the objector, who was pleased to contend that the gaseous nebula was a figment of the imagination, could never have been effectively silenced by Kant or Laplace or Herschel. It would have been useless for them to point to the Nebula in Orion, for the objector might say that it was only a cluster of stars, and at that time there would have been no way of confuting him.

The authors of the Nebular Theory had, in respect to this class of objector, a much more difficult task than falls to its modern advocate. The latter is able to deny in the most emphatic manner that a gaseous nebula is no more than an imaginary conception.

The famous discovery of Sir W. Huggins has removed the first great objection to the Nebular Theory.

Chapter 5

THE HEAT OF THE SUN

It will be convenient to consider different bodies in the solar system, and to study them with the special object of ascertaining what information they afford as to the great celestial evolution. We cannot hesitate as to which of the bodies should first claim our attention. Not on account of the predominant importance of our sun to the inhabitants of the earth, but rather because the sun is nearly a thousand times greater than the greatest of the planets, do we assign to the great luminary the first position in this discussion.

The sun is, indeed, especially instructive on the subject with which we are occupied. By reason of its great mass, the process of evolution takes place more slowly in the sun than in the earth or in any other planet. Evolution has, no doubt, largely transformed the sun from its primæval condition, but it has not yet produced a transformation so radical as that which the earth and the other planets have undergone. On this account the sun can give us information about the process of evolution which is not to be so easily obtained from any of the other heavenly bodies. The sun can still exhibit to us some vestiges, if we may so speak, of that great primæval nebula from which the whole system has sprung.

The heat of the sun is indeed one of the most astonishing conceptions which the study of Nature offers to us. Let me try to illustrate it. Think first of a perfect modern furnace in which even steel itself, having first attained a dazzling brilliance, can be further melted into a liquid that will run like water. Let us imagine the temperature of that liquid to be multiplied seven-fold, and then we shall obtain some conception of the fearful intensity of the heat which would be found in that wonderful celestial furnace the great sun in the heavens.

Ponder also upon the stupendous size of that orb, which glows at every point of its surface with the astonishing fervour that this illustration suggests. The earth on which we stand is a mighty globe; yet what are the dimensions of our earth in comparison with those of the sun? If we represent the earth by a grain of mustard seed, then on the same scale the sun should be represented by a cocoanut. We may perhaps obtain a more impressive conception of the proportions of the orb of day in the following manner. Look up at the moon which revolves round the heaven, describing as it does so majestic a track that it is generally at a distance of two hundred and forty thousand miles from the earth. Yet the sun is so large that if there were a hollow globe equally great, and the earth were placed at its centre, the entire orbit of the moon would lie completely within it.

Every portion of that stupendous desert of flame is pouring forth torrents of heat. It has, indeed, been estimated that the heat which issues from an area of two square feet on the sun would more than suffice, if it could be all utilised, to drive the engines of the largest Atlantic liner between Liverpool and New York.

This solar heat is scattered through space with boundless prodigality. No doubt the dwellers on the earth do receive a fair supply of sunbeams; but what is available for the use of mankind can be hardly more than an infinitesimal fraction of what the sun emits. We shall scarcely be so presumptuous as to suppose that the sun has been designed solely for the benefit of the poor humanity which needs light and warmth. The heat and light daily lavished by the sun would suffice to warm and to illuminate two thousand million globes, each as great as the earth. If, indeed, it were true that the only object of the sun's existence was to cherish this immediate world of ours, then all we can say is that the sun carries on its business in a most outrageously wasteful manner. What would be thought of the prudence of one who, having been endowed with a fortune of ten million pounds, spent one single penny of that vast sum in a profitable manner and dissipated every other penny and every other pound of his fortune in aimless extravagance? But this is apparently the way in which the sun manages its affairs, so far as our earth is concerned. Out of every ten million pounds worth of heat issuing from the glorious orb of day, we on this earth secure one pennyworth, and all but that solitary pennyworth seems to be utterly squandered. We may say it certainly is squandered so far as humanity is concerned. What, indeed, its actual destination may be science is unable to tell.

And now for the great question as to how the sun's heat is sustained. How is it that this career of tremendous prodigality has not ages ago been checked by absolute exhaustion? Every child knows that the fire on the hearth will go out unless coal be provided. The workman knows that his devouring furnace in the ironworks requires to be incessantly stoked with fresh supplies of fuel. How, then, comes it that the wonderful furnace on high can still continue, as it has continued for ages, to pour forth its amazing stores of heat without being exhausted?

Professor Langley has supplied us with an admirable illustration showing the amount of fuel which would be necessary, if indeed it were by successive additions of fuel that the sun's heat was sustained. Suppose that all the coal-seams which underlie England and Scotland were made to yield up their stores; that the vast coalfields in America, Australia, China, and elsewhere were compelled to contribute every combustible particle they contained; suppose, in fact, that we extracted from this earth every ton of coal which it possesses in every isle and every continent; suppose that this mighty store of fuel, sufficient to supply all the wants of the earth for centuries, were to be accumulated, and that by some mighty effort that mass were to be hurled into the sun and were forthwith to be burnt to ashes; there can be no doubt that a stupendous quantity of heat would be produced. But what is that heat in comparison, we do not say with the heat of the sun, but with the daily expenditure of the sun's heat? How long, think you, would the combustion of so vast a mass of fuel provide for the sun's expenditure? We are giving deliberate expression to a scientific fact when we say that a conflagration which destroyed every particle of coal contained in this earth would not generate as much heat as the sun lavishes in the tenth part of every single second. During the few minutes that you have been reading these words a quantity of heat has gone forever from the sun which is five thousand times as great as all the heat that ever has been or ever will be produced by the combustion of the coal that this earth has furnished.

But we have still another conception to introduce before we can appreciate the full significance of the sun's extraordinary expenditure of heat and light. We have been thinking of the sun as it shines now; but as the sun shines to-day, so it has shone yesterday, and so it shone a hundred years ago, a thousand years ago; so it shone in the earliest dawn of history, so it shone during those still remoter periods when great animals flourished which have now vanished forever; so the sun shone during those remote ages when life began to dawn on an earth which still was young. We do not, indeed, say that the intensity of the sunbeams has remained actually uniform throughout a period so vast; but there is every reason to believe that throughout these illimitable periods the sun has expended its radiance with the most lavish generosity.

A most important question is suggested by these considerations. The consequences of frightful extravagance are known to us all; we know that such conduct tends to bankruptcy and ruin; and certainly the expenditure of heat by the sun is the most magnificent extravagance of which our knowledge gives us any conception. Accordingly, the important question arises: As to how the consequences of such awful prodigality have been hitherto averted. How is it that the sun is still able to draw on its heat reserve, from year to year, from century to century, from æon to æon, ever squandering two thousand million times as much heat as that which genially warms our temperate regions, as that which draws forth the exuberant vegetation of the tropics or which rages in the desert of Sahara? That is the great problem to which our attention has to be given.

We must first ascertain, with such precision as the circumstances permit, the actual amount of heat which the sun pours forth in its daily radiation. The determination of this quantity has engaged the attention of many investigators, and the interpretation of their results is by no means free from difficulty. It is to be observed that what we are now seeking to ascertain is not exactly a question of temperature, but of something quite different. What we have to measure is a quantity of heat, which is to be expressed in the proper units for quantities of heat. The unit of heat which we shall employ is the quantity of heat necessary to raise one pound of water through one degree Fahrenheit.

The *solar constant* is the number of units of heat which fall, in one minute, on one square foot of a surface placed at right angles to the sun's rays, and situated at the mean distance of the earth from the sun. We shall suppose that losses due to atmospheric absorption have been allowed for, so that the result will express the number of units of heat that would be received in one minute on a square foot turned directly to the sun, and at a distance of 93,000,000 miles.

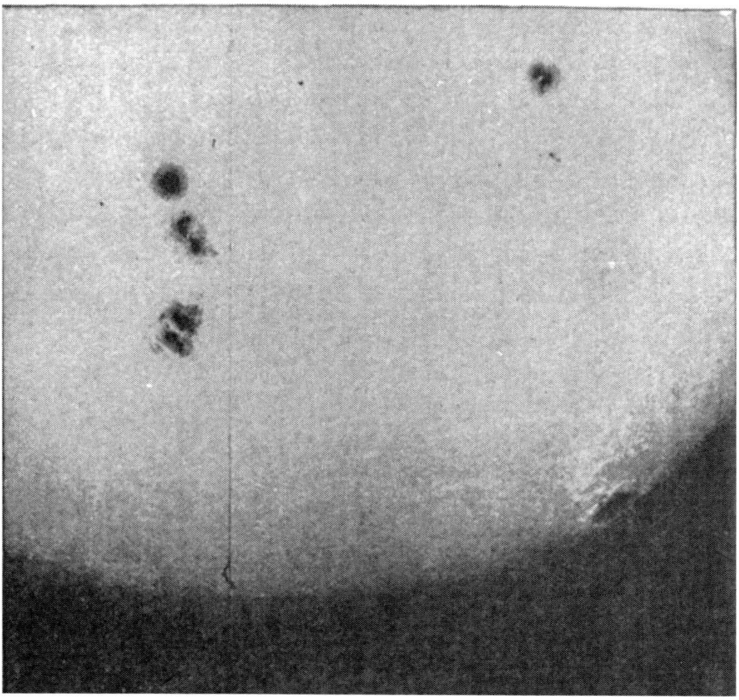

Figure 15. The Sun (July 8th, 1892). (Royal Observatory, Greenwich.)
(*From the Royal Astronomical Society Series.*)

This is a matter for determination by actual observation and measurement. Theory can do little more than suggest the precautions to be observed and discuss the actual figures which are obtained. There have been many different methods of making the observations, and the results are somewhat various, but the discrepancies are not greater than might be expected in an investigation of such difficulty. The mean value which has been arrived at is *fourteen*, and the fundamental fact with regard to the solar radiation which we are thus enabled to state is that an area of a square foot exposed at right angles to the solar rays, at a distance of 93 millions of miles, will in each minute receive from the sun as much heat as would raise one pound of water fourteen degrees Fahrenheit.

It follows that the total radiation from the sun must suffice to convey, in each minute, to the surface of a sphere whose radius is 93,000,000 miles, fourteen units of heat per square foot of that surface. This radiation comes from the surface of the sun. It is easily shown that the heat from each square foot on the sun will have to supply an area of 46,000 square feet at the distance of the earth. Hence the number of units of heat emerging each minute from a square foot on the sun's surface must be about 640,000.

We can best realise what this statement implies by finding the amount of coal which would produce the same quantity of heat. It can be shown that the heat given out by each square foot of the solar surface in one minute will be equivalent to that produced in the combustion of forty-six pounds of coal. If the sun's heat were sustained by combustion, every part of the sun's surface as large as the grate of an ordinary furnace would have to be doing at least one hundred times as much heating as the most vigorous stoking could extract from any actual furnace.

The radiation of heat from a single square foot of the solar surface in the course of a year must, therefore, be equivalent to the heat generated in the combustion of 11,000 tons of the best coal. If we estimate the annual coal production of Great Britain at 250,000,000 tons, we find that the total heat which this coal can produce is not greater than the annual emission from a square of the sun's surface of which each side is fifty yards. All the coal exported from England in a year does not give as much heat as the sun radiates in the same time from every patch on its surface which is as big as a croquet ground.

There is perhaps no greater question in the study of Nature than that which enquires how the sun's heat is sustained so that the radiation is still dispensed with unstinted liberality. If we are asked how the sun can be fed so as to sustain this expenditure, we have to explain that the sun is not really fed. If, then, it receives no adequate supplies of energy from without, we have to admit that the sun must be getting exhausted.

I ought, indeed, to anticipate objection by at once making the admission that the sun does receive some small supply of energy from the meteors which are perennially drawn into it. The quantity of energy they yield is, however, insignificant in comparison with the solar expenditure of heat. We may return to this subject at a later period, and it need not now receive further attention.

We must deliberately face the fact that the energy of the sun is becoming exhausted. But the rate of exhaustion is so slow that it affords no prospect of inconvenience to humanity; it does not excite alarm. We grant that we are not able to observe by instrumental means any perceptible diminution of solar energy. Still, as we know that energy is being steadily dissipated from the sun, and that energy cannot be created from nothing, it is certain the decline is in progress. But the reserve of energy which the sun possesses, and which can be ultimately rendered available to sustain the radiation, is so enormous in comparison with the annual expenditure of energy, that myriads of centuries will have to elapse before there is any appreciable alteration in the effectiveness of the sun.

Let me illustrate the point by likening the sun to a grain warehouse, in which 2,500 tons of wheat can be accommodated. Let us suppose that the warehouse was quite full at the beginning, and that the wheat was to be gradually abstracted, but only at the rate of one grain each day. Let us further suppose that no more wheat is to be added to that already in the warehouse, and let us assume that the wheat thus stored away experiences no deterioration and no loss whatever except by the removal of one grain per diem. It is easy to see that very many centuries would have to elapse before the grain in that warehouse had decreased to any appreciable extent.

With a consumption at the rate of a single grain a day a ton of corn would last about four thousand years, and 2,500 tons of corn would accordingly last about ten million years. It follows, therefore, that if the grain in that store were consumed at the rate of only one grain per day the warehouse would not be emptied for ten million years.

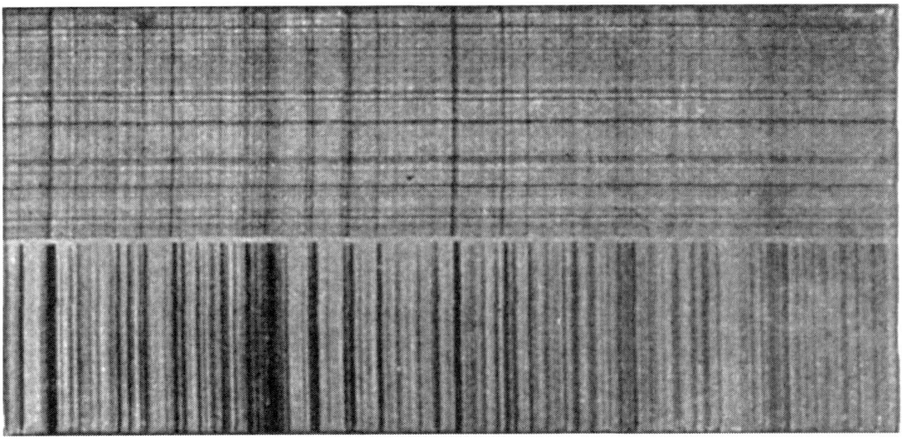

I.

II.

Figure 16. I. Spectrum of the Sun. II. Spectrum of Arcturus.
(*Professor H. C. Lord.*)

The quantity of heat, or rather the reserve of energy equivalent to heat, which still remains stored up in the sun bears to the quantity of heat which the sun radiates away in a single day a ratio something like that which a single grain of corn bears to 2,500 tons of corn.

The sun's potential store of heat is no doubt very great, though not indefinitely great. That heat is beyond all doubt to be ultimately exhausted; but the reserve is so prodigious that for the myriads of years during which the sun has been subjected to human observation there has been no appreciable alteration in the efficiency of radiation.

It might be supposed that the sun was merely a white-hot globe cooling down, and that the solar radiation was to be explained in this way. But a little calculation will prove it to be utterly impossible that the heat of the great luminary could be so accounted for. A knowledge of the current expenditure of solar heat shows that if the sun had been a globe of iron at its fusing point, then at the present rate of radiation it would have sunk to the temperature of freezing water in forty-eight years.

Perhaps I ought here to explain a point which might otherwise cause misapprehension. For our ordinary sources of artificial heat we, of course, employ some form of combustion. Whenever combustion takes place there is chemical union between the carbon or other fuel, whatever it may be, and the oxygen of the atmosphere. A certain quantity of carbon enters into chemical union with a definite quantity of oxygen, and, as an incident in the process, a definite quantity of heat is liberated. So much coal, for instance, requires for complete combustion so much air, and, granted a sufficiency of air, the union of the carbon and hydrogen in the coal will give out a certain quantity of heat which may be conveniently expressed by the number of pounds of water which that heat would suffice to transform into steam. It is necessary to observe that there are definite numerical relations among these quantities. The quantity of heat that can be produced by the combustion of a pound of any particular substance will depend upon the nature of that substance.

As chemical combination is the main source of the artificial heat which we employ for innumerable purposes on the earth, it seems proper to consider whether it can be any form of chemical combination which constitutes the source of the heat which the sun radiates in such abundance. It is easy to show that the solar radiation cannot be thus sustained. The point to which I am now referring was very clearly illustrated by Helmholtz in a lecture he delivered many years ago on the origin of the planetary system.

To investigate whether the solar heat can be attributed to chemical combination, we shall assume for the moment that the sun is composed of those particular materials which would produce the utmost quantity of heat for a given weight; in other words, that the sun is formed of hydrogen and oxygen in quantities having the same ratio as that in which they should be united to form water. The quantity of heat generated by the union of known weights of oxygen and hydrogen has been ascertained, by experiments in the laboratory, to exceed that which can be generated by corresponding weights of any other materials. We can calculate how much of the sun's mass, if thus constituted, would have to enter into combination every hour in order to generate as much heat as the hourly radiation of the sun. We need not here perform the actual calculation, but merely state the result, which is a very remarkable one. It shows that the heat arising from the supposed chemical action would not suffice to sustain the radiation of the sun at its present rate for more than 3,000 years. Thirty centuries is a long time, no doubt, yet still we must remember that it is no more than a part even of the period known to human history. If, indeed, it had been by combustion that the sun's heat was produced, then from the beginning of the sun's career as a luminous object to its final extinction and death could not be longer than 3,000 years, if we assumed that its radiation was to be uniformly that which it now dispenses.

But it may be said that we are dealing only with elements known to us and with which terrestrial chemists are familiar, and it may be urged that the sun possibly contains materials whose chemical union produces heat in much greater abundance than do the elements with which alone we are acquainted. But this argument cannot be sustained. One of the most important discoveries of the last century, the discovery which perhaps more than any other has tended to place the nebular theory in an impregnable position, is that which tells us that the elements of which the sun is composed are the same as the elements of which our earth is made. We shall have to refer to this in detail in a later chapter. We now only make this passing reference to it in order to dismiss the notion that there can be unknown substances in the sun whose heat of combustion would be sufficiently great to offer an explanation of the extraordinary abundance of solar radiation.

There is nothing more characteristic of the physical science of the century just closed than the famous discovery of the numerical relation which exists between heat and energy. We are indebted to the life-long labours of Joule, followed by those of many other investigators, for the accurate determination of the fundamental constant which is known as the mechanical equivalent of heat. Joule showed that the quantity of heat which would suffice to raise one pound of water through a single degree Fahrenheit was the precise equivalent of the quantity of energy which would suffice to raise 772 pounds through a height of one foot. It would be hard to say whether this remarkable principle has had a more profound effect on practical engineering or on the course of physical science. In practical engineering, the knowledge of the mechanical equivalent of heat will show the engineer the utmost amount of work that could by any conceivable apparatus be extracted from the heat potentially contained in a ton of coal. In the study of astronomy the application of the same principle will suffice to explain how the sun's heat has been sustained for illimitable ages.

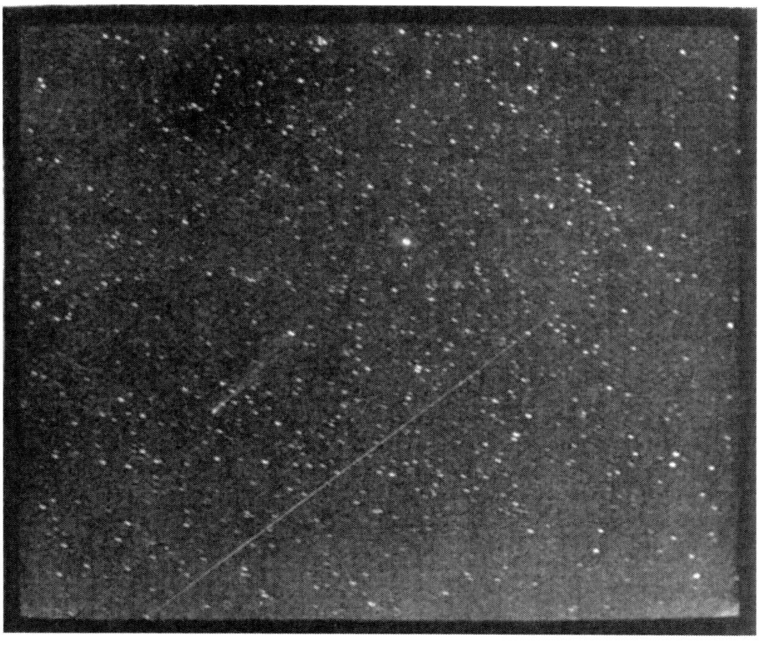

Figure 17. Brooks' Comet and Meteor Trail. (November 13th, 1893. Exposure 2 hours.) (*Photographed by Professor E. E. Barnard.*)

It will be convenient to commence with a little calculation, which will provide us with a result very instructive when considering celestial phenomena in connection with energy. We have seen that the unit of heat—for so we term the quantity of heat necessary to raise a pound of water one degree—will suffice, when transformed into mechanical energy, to raise 772 pounds through a single foot. This would, of course, be precisely the same thing as to raise one pound through 772 feet. Suppose a pound weight were carried up 772 feet high and were then allowed to drop. The pound weight would gradually gather speed in its descent, and, at the moment when it was just reaching the earth, would be moving with a speed of about 224 feet a second. We may observe that the work which was done in raising the body to this height has been entirely expended in giving the body this particular velocity. A weight of one pound, moving with a speed of 224 feet a second, will therefore contain, in virtue of that motion, a quantity of energy precisely equivalent to the unit of heat.

It is a well-known principle in mechanics that if a body be dropped from any height, the velocity with which it would reach the ground is just the velocity with which the body should be projected upwards from the ground in order to re-ascend to the height from which it fell (the resistance of the air is here overlooked as not having any bearing upon the present argument). Thus we see that a weight, moving with a velocity of 224 feet per second, contains within itself, in virtue of its motion, energy adequate to make it ascend against gravity to the height of 772 feet. That is to say, this velocity in a body of a pound weight can do for the body precisely what the unit of heat can do for it; hence we say that in virtue of its movement the body contains a quantity of energy equal to the energy in the unit of heat.

Let us now carry our calculation a little further. If a pound of good coal be burned with a sufficient supply of oxygen, and if every precaution be taken so that no portion of the heat be wasted, it can be shown that the combustion of the coal is sufficient to produce 14,000 units of heat. In other words, the burning of one pound of coal ought to be able to raise 14,000 pounds of water one degree, or 140 pounds of water a hundred degrees, or 70 pounds of water two hundred degrees. I do not mean to say that efficiency like this will be attained in the actual circumstances of the combustion of coal in the fireplace. A pound of coal does, no doubt, contain sufficient heat to boil seven gallons of water; but it cannot be made to effect this, because the fireplace wastes in the most extravagant manner the heat which the coal produces, so that no more than a small fraction of that heat is generally rendered available. But in the cosmical operations with which we shall be concerned we consider the full efficiency of the heat; and so we take for the pound of coal its full theoretical equivalent, namely, 14,000 thermal units. Let us now find the quantity of energy expressed in foot-pounds[2] to which this will correspond. It is obtained by multiplying 14,000 units of heat by 772, and we get as the result 10,808,000. That is to say, a pound of good coal, in virtue of the fact that it is combustible and will give out heat, contains a quantity of energy which is represented by ten or eleven million foot-pounds.

We now approach the question in another way. Let us think of a piece of coal in rapid motion; if the coal weighed a pound, and if it were moving at 224 feet a second, then the energy it contains in consequence of that velocity would, as we have seen, correspond to one thermal unit. We have, however, to suppose that the piece of coal is moving with a speed much higher than that just stated; and here we should note that the energy which a moving body possesses, in virtue of its velocity, increases very rapidly when the speed of that body increases. If the velocity of a moving body be doubled, the energy that it possesses increases fourfold. If the velocity of the body be increased tenfold, then the energy that it possesses will be increased a hundredfold. More generally, we may say that the energy of a moving body is proportional to the square of the velocity with which the body is animated. Let us, then, suppose that the piece of coal, weighing one pound, is moving with a speed as swift as a shot from the finest piece of artillery, that is to say, with a speed of 2,240 feet a second; and as this figure is ten times 224, it shows us that the moving body will then possess, in virtue of its velocity, the equivalent of one hundred units of heat.

But we have to suppose a motion a good deal more rapid than that obtained by any artillery; we shall consider a speed rather more than ten times as fast. It is easy to calculate that if the piece of coal which weighs a pound is moving at the speed of five miles a second, the energy that it would possess in consequence of that motion would approximate to 14,000 thermal units. In other words, we come to the conclusion that any body moving with a velocity of five miles a second will possess, in virtue of that

[2] A foot-pound is the amount of energy required to raise a pound weight through a height of one foot.

velocity, a quantity of energy just equal to the energy which an equally heavy piece of good coal could produce if burnt in oxygen, and if every portion of the heat were utilised.

It is quite true that the speed of five miles a second here supposed represents a velocity much in excess of any velocity with which we are acquainted in the course of ordinary experience. It is more than ten times as fast as the speed of a rifle bullet. But a velocity of five miles a second is not at all large when we consider the velocities of celestial bodies. We want this fact relating to the energy in a piece of coal to be remembered. We shall find it very instructive as our subject develops, and therefore we give some illustrations with reference to it.

The speed of the earth as it moves round the sun is more than eighteen miles a second—that is to say, it is three and a half times the critical speed of five miles. In virtue of this speed the earth has a corresponding quantity of energy. To find the equivalent of that energy it must, as already explained, be remembered that the energy of a moving body is proportional to the square of its velocity; it follows that the energy of the earth, due to its motion round the sun, must be almost twelve times as great as the energy of the earth would be if it moved at the rate of only five miles a second. But, we have already seen that a body with the velocity of five miles a second would, in virtue of that motion, be endowed with a quantity of energy equal to that which would be given out by the perfect combustion of an equal weight of coal. It follows, therefore, that this earth of ours, solely in consequence of the fact that it is moving in its orbit round the sun, is endowed with a quantity of energy twelve times as great as all the energy that would be given out in the combustion of a mass of coal equal to the earth in weight. This may seem an astonishing statement; but its truth is undoubted. If it should happen that the earth came into collision with another body by which its velocity was stopped, the principle of the conservation of energy tells us that this energy, which the earth has in consequence of its motion, must forthwith be transformed, and the form which it will assume is that of heat. Such a collision would generate as much heat as could be produced by the combustion of twelve globes of solid coal, each as heavy as the earth. We may indeed remark that the coal-seams in our earth's crust contain, in virtue of the fact that they partake of the earth's orbital motion, twelve times as much energy as will ever be produced by their combustion.

It can hardly be doubted that such collisions as we have here imagined do occasionally happen in some parts of space. Those remarkable new stars which from time to time break out derive, in all probability, their temporary lustre from collisions between bodies which were previously non-luminous. But we need not go so far as inter-stellar space for a striking illustration of the transformation of energy into heat. In the pleasing phenomena of shooting stars our own atmosphere provides us with beautiful illustrations of the same principle. The shooting star so happily caught on Professor Barnard's plate may be cited as an example.

Chapter 6

HOW THE SUN'S HEAT IS MAINTAINED

The law which declares that a body which gives out heat must in general submit to a corresponding diminution in volume appears, so far as we can judge, to be one of those laws which have to be obeyed not alone by bodies on which we can experiment, but by bodies throughout the extent of the universe. The law which bids the mercury ascend the stem of the thermometer when the temperature rises, and descend when the temperature falls, affords the principle which explains some of the grandest phenomena of the heavens. Applied to the solar system it declares that as the sun, in dispensing its benefits to the earth day by day, has to pour forth heat, so in like manner must it be diminishing in bulk.

Assuming that this principle extends sufficiently widely through time and space, we shall venture to apply its consequences over the mighty spaces and periods required for celestial evolution. We disdain to notice the paltry centuries or mere thousands of years which include that infinitesimal trifle known as human history. Our time conceptions must undergo a vast extension.

It was Helmholtz who first explained by what agency the sun is able to continue its wonderful radiation of heat, notwithstanding that it receives no appreciable aid from chemical combination. Helmholtz pointed out that inasmuch as the sun is pouring out heat it must, like every other cooling body, contract. We ought not, indeed, to say every cooling body; it would be more correct to say, everybody which is giving out heat, for the two things are not necessarily the same. Indeed, strange as it may appear, it would be quite possible that a mass of gas should be gaining in temperature even though it were losing heat all the time. At first this seems a paradox, but the paradox will be explained if we reflect upon the physical changes which the gas undergoes in consequence of its contraction.

Let us dwell for a moment on the remarkable statement that the sun is becoming gradually smaller. The reduction required to sustain the radiation corresponds to a diminution of the diameter by about a mile every eleven years. It may serve to impress upon us the fact of the sun's shrinkage if we will remember that on that auspicious day when Queen Victoria came to the throne the sun had a diameter more than five miles greater than it had at the time when her long and glorious career was ended. The sun that shone on Palestine at the beginning of the present era must have had a diameter about one hundred and seventy miles greater than the sun which now shines on the Sea of Galilee. This process of reduction has been going on for ages, which from the human point of view we may practically describe as illimitable. The alteration in the sun's diameter within the period covered by the records of man's sway on this earth may be intrinsically large; it amounts no doubt to several hundreds of miles. But in comparison with the vast bulk of the sun this change in its magnitude is unimportant. A span of ten thousand years will certainly include all human history. Let us take a period which is four times as long. It is easy to calculate what the diameter of the sun must have been forty thousand years ago, or what the diameter of the sun is to become in the next forty thousand years. Calculated at the rate we have given, the alteration in the sun's diameter in this period amounts to rather less than four thousand miles. This seems no doubt a huge alteration in the dimensions of the orb of day. We must, however, remember that at the present moment the diameter of the sun is about 863,000 miles, and that a loss of four thousand miles, or thereabouts, would still leave a sun with a diameter of 859,000 miles. There would not be much recognisable difference between two suns of these different dimensions. I think I may say that if we could imagine two suns in the sky at the same moment, which differed only in the circumstance that one had a diameter of 863,000 miles and the other a diameter of 859,000 miles, it would not be possible without careful telescopic measurement to tell which of the two was the larger.

After a contraction has taken place by loss of heat, the heat that still remains in the body is contained within a smaller volume than it had originally. The temperature depends not only on the actual quantity of heat that the mass of gas contains, but also on the volume through which that quantity of heat is diffused. If there be two equal weights of gas, and if they each have the same absolute quantity of heat, but if one of them occupies a larger volume than the other, then the temperature of the gas in the large volume will not be so high as the temperature of the gas in the smaller volume. This is indeed so much the case, that the reduction of volume by the loss of heat may sometimes have a greater effect in raising the temperature than the very loss of heat which produced the contraction has in depressing it. On the whole, therefore, a gain of temperature may be shown. This is what, indeed, happens not unfrequently in celestial bodies. The contraction having taken place, the lesser quantity of heat still shows to such advantage in the reduced volume of the body, that no decline of temperature will be perceptible. It may happen that simultaneously with the decrease of heat there is even an increase of temperature.

The principle under consideration shows that, though the sun is now giving out heat copiously, it does not necessarily follow that it must at the same time be sinking in temperature. As a matter of fact, physicists do not know what course the temperature of the sun is actually taking at this moment. The sun may now be precisely at the same temperature at which it stood a thousand years ago, or it may be cooler, or it may be hotter. In any case it is certain that the change of temperature per century is small, too small, in fact, to be decided in the present state of our knowledge. We cannot observe any change, and to estimate the change from mechanical principles would only be possible if we knew much more about the interior of the sun than we know at present.

We are forced to the conclusion that the energy of the sun, by which we mean either its actual heat or what is equivalent to heat, must be continually wasting. A thousand years ago there was more heat, or its equivalent, in the sun than there is at present. But the sun of a thousand years ago was larger than the sun that we now have, and the heat, or its equivalent, a thousand years ago may not have been so effective in sustaining the temperature of the bigger sun as the lesser quantity of heat is in sustaining the temperature of the sun at the present day. It will be noticed that the argument depends essentially on the alteration of the size of the sun. Of course if the orb of day had been no greater a thousand years ago than it is now, then the sun of those early days would not only have contained more heat than our present sun, but it must have shown that it did contain more heat. In other words, its temperature would then certainly have been greater than it is at present.

Thus we see the importance—so far as radiation is concerned—of the gradual shrinking of the sun. The great orb of day decreases, and its decrease has been estimated numerically. We cannot, indeed, determine the rate of decrease by actual telescopic measurement of the sun's disc with the micrometer; observations extending over a period of thousands of years would be required for this purpose. But from knowing the daily expenditure of heat from the sun it is possible to calculate the amount by which it shrinks. We cannot conveniently explain the matter fully in these pages. Those who desire to see the calculation will find it in the Appendix. Suffice it to say here that the sun's diameter diminishes about sixteen inches in every twenty-four hours. This is an important conclusion, for the rate of contraction of the solar diameter is one of the most significant magnitudes relating to the solar system.

It was Helmholtz who showed that the contraction of the sun's diameter by sixteen inches a day is sufficient to account for the sustentation of the solar radiation. For immense periods of time the heat may be dispensed with practically unaltered liberality. The question then arises as to what time-limit may be assigned to the efficiency of our orb. Obviously the sun cannot go on contracting sixteen inches a day indefinitely. If that were the case, a certain number of millions of years would see it vanish altogether. The limit to the capacity of the sun to act as a dispenser of light and heat can be easily indicated. At present the sun, in its outer parts at all events, is strictly a vaporous body. The telescope shows us nothing resembling a solid or a liquid globe. The sun seems composed of gas in which clouds and vapours are suspended. In the sun's centre the temperature is probably very much greater than any temperature which can be produced by artificial means; it would doubtless be sufficient not only to melt, but even to drive into vapour the most refractory materials. On the other hand, the enormous condensing pressure to which those materials are submitted by the stupendous mass of the sun will have the effect of keeping them together and of compressing them to such an extent that the density of the gas, if indeed we may call it gas, is probably as great as the density of any known matter. The fact is that the terms liquids, gases, and solids cease to retain intelligible distinctions when applied to materials under such pressure as would be found in the interior of the sun.

Figure 18. Argo and the Surrounding Stars and Nebulosity.
(*Photographed by Sir David Gill, K.C.B.*)

Astronomers can weigh the sun. It may well be imagined that this is a delicate and difficult operation. It can, however, be effected with but little margin of uncertainty, and the result is a striking one. It serves no useful purpose to express the sun's weight as so many myriads of tons. It is more useful for our present purpose to set down the density of the sun, that is to say, the ratio of the weight of the orb, to that of a globe of water of the same size. This is the useful form in which to consider the weight of the sun. Astronomers are accustomed to think of the weight of our own earth in this same fashion, and the result shows that the earth is rather more than five times as heavy as a globe of water of the same size. We can best appreciate this by stating that if the earth were made of granite, and had throughout the density which we find granite to possess at the surface, our globe would be about three times as heavy as a globe of water of the same size. If, however, the earth had been entirely made of iron, it would be more than seven times as heavy as a globe of water of the same size. As the earth actually has a density of 5, it follows that our globe taken as a whole is heavier than a globe of granite of the same size, though not so heavy as a globe of iron.

In the matter of density there is a remarkable contrast between the sun and the earth. The sun's density is much less than that of the earth. Of course it will be understood that the sun is actually very much heavier than our globe; it is indeed more than three hundred thousand times greater in weight. But the sun is about a million three hundred thousand times as big as the earth, and it follows from these figures that its density cannot be more than about a fourth of that of the earth. The result is that, at present, the sun is nearly half as heavy again as a globe of water the same size. We have used round numbers: the density of the sun is actually 1.4.

In the following manner we explain how heat is evolved in the contraction of the sun. In its early days the sun, or rather the materials which in their aggregate form now constitute the sun, were spread over an immense tract of space, millions of times greater than the present bulk of the sun. We see nebulosities even now in the heavens which may suggest what the primæval nebula may have been before the evolution had made much progress. Look for instance at Sir David Gill's photograph of the

Nebula in Argo, or at the Trifid Nebula. We may, indeed, consider the primæval nebula to have been so vast that particles from the outside falling into the position of the present solar surface would acquire that velocity of three hundred and ninety miles a second which we know the attraction of the sun is capable of producing on an object which has fallen in from an indefinitely great distance. As these parts are gradually falling together at the centre, there will be an enormous quantity of heat developed from their concurrence. Supposing, for instance, that the materials of the sun were arranged in concentric spherical shells around the centre, and imagining these shells to be separated by long intervals, so that the whole material of the sun would be thus diffused over a vast extent, then every pound weight in the outermost shell, by the very fact of its sinking downwards to the present solar system, would acquire a speed of 390 miles a second, and this corresponds to as much energy as could be produced by the burning of three tons of coal. But be the fall ever so gentle, the great law of the conservation of energy tells us that for the same descent, however performed, the same quantity of heat must be given out. Each pound in the outer shell would therefore give out as much heat as three tons of coal. Every pound in the other shells, by gradual descent into the interior, would also render its corresponding contribution. It then becomes easily intelligible how, in consequence of the original diffusion of the materials of the sun over millions of times its present volume, a vast quantity of energy was available. As the sun contracted this energy was turned into radiant heat.

We may anticipate a future chapter so far as to assume that there was a time when even this solid earth of ours was a nebulous mass diffused through space. We are not concerned as to what the temperature of that nebulous mass may have been. We may suppose it to be any temperature we please. The point that we have now to consider is the quantity of heat which is generated by the contraction of the nebula. That heat is produced in the contraction will be plain from what has gone before. But we may also demonstrate it in a slightly different way. Let us take any two points in the nebula, P and Q. After the nebula has contracted the points which were originally at P and Q will be found at two other points, A and B. As the whole nebula in its original form was larger than the nebula after it has undergone its contraction, the distance P Q is generally greater than the distance A B. We may suppose the contraction to proceed uniformly, so that the same will be true of the distance between any other two particles. The distance between every pair of particles in the contracted nebula will be less than the distance between the same particles in the original nebula.

Figure 19. Trifid Nebula in Sagittarius (Lick Observatory, California). (*From the Royal Astronomical Society Series.*)

If two attracting bodies, A and B, are to be moved further apart than they were originally, force must be applied and work must be done. We may measure the amount of that work in foot-pounds, and then, remembering that 772 foot-pounds of work are equivalent to the unit of heat, we may express the energy necessary to force the two particles to a greater distance asunder in the equivalent quantity of heat. If, therefore, we had to restore the nebula from the contracted state to the original state, this would involve a forcible enlargement of the distance A B between every two particles to its original value, P Q. Work would be required to do this in every case, and that work might, as we have explained, be expressed in terms of its equivalent heat value. Even though the temperature of the nebula is the same in its contracted state as in its original state, we see that a quantity of heat might be absorbed or rendered latent in forcing the nebula from one condition to the other. In other words, keeping the temperature of the nebula always constant, we should have to apply a large quantity of heat to change the nebula from its contracted form to its expanded form.

It is equally true that when the nebula is contracting, and when the distance between every two particles is lessening, the nebula must be giving out energy, because the total energy in the contracted state is less than it was in the expanded state. This energy is equivalent to heat. We need not here pause to consider by what actual process the heat is manifested; it suffices to say that the heat must, by one of the general laws of Nature, be produced in some form.

We are now able to make a numerical estimate. We shall suppose that the earth, or rather the materials which make the earth, existed originally as a large nebula distributed through illimitable space. The calculations show that the quantity of heat, generated by the condensation of those materials from their nebulous form into the condition which the earth now has, was enormously great. We need not express this quantity of heat in ordinary units. The unit we shall take is one more suited to the other dimensions involved. Let us suppose a globe of water as heavy as the earth. This globe would have to be five or six times as large as the earth. Next let us realise the quantity of heat that would be required to raise that globe of water from freezing point to boiling point. It can be proved that the heat, or its equivalent, which would be generated merely by the contraction of the nebula to form the earth, would be ninety times as great as the amount of heat which would suffice to raise a mass of water equal in weight to the earth from freezing point to boiling point.

We apply similar calculations to the case of the sun. Let us suppose that the great luminary was once diffused as a nebula over an exceedingly great area of space. It might at first be thought that the figures we have just given would answer the question. We might perhaps conjecture that the quantity of heat would be such as would raise a mass of water equal to the sun's mass from freezing to boiling point ninety times over. But we should be very wrong in such a determination. The heat that is given out by the sun's contraction is enormously greater than this estimate would represent, and we shall be prepared to admit this if we reflect on the following circumstances. A stone falling from an indefinitely great distance to the sun would acquire a speed of 390 miles a second by the time it reached the sun's surface. A stone falling from an indefinitely great distance in space to the earth's surface would, however, acquire a speed of not more than seven miles a second. The speed acquired by a body falling into the sun by the gravitation of the sun is, therefore, fifty-six times as great as the speed acquired by a body falling from infinity to the earth by the gravitation of the earth. As the energy of a moving body is proportional to the square of its velocity, we see that the energy with which the falling body would strike the sun, and the heat that it might consequently give forth, would be about three thousand times as great as the heat which would be the result of the fall of that body to the earth. We need not therefore be surprised that the drawing together of the elements to form the sun should be accompanied by the evolution of a quantity of heat which is enormously greater than the mere ratio of the masses of the earth and sun would have suggested.

There is another line of reasoning by which we may also illustrate the same important principle. Owing to the immense attraction possessed by the large mass of the sun, the weights of objects on that luminary would be very much greater than the weights of corresponding objects here. Indeed, a pound on the sun would be found by a spring-balance to weigh as much as twenty-seven pounds here. If the materials of the sun had to be distributed through space, each pound lifted a foot would require twenty-seven times the amount of work which would be necessary to lift a pound through a foot on the earth's surface. It will thus be seen that not only the quantity of material that would havc to be displaced is enormously greater in the sun than in the earth, but that the actual energy that would have to be applied per unit of mass from the sun would be many times as great as the quantity of energy that would have to be applied per unit of mass from the earth to effect a displacement through the same distance. To distribute the sun's materials into a nebula we should therefore require the expenditure of a quantity of work far more than proportional to the mere mass of the sun. It follows that when the sun is contracting the quantity of work that it will give out, or, what comes to the same thing, the amount of heat that would be poured forth in consequence of the contraction per unit of mass of the sun will largely exceed the quantity of heat given out in the similar contraction of the earth per unit of mass of the earth.

These considerations will prepare us to accept the result given by accurate calculation. It has been shown that the heat which would be generated by the condensation of the sun from a nebula filling all space down to its present bulk is two hundred and seventy thousand times the amount of heat which would be required to raise the temperature of a mass of water equal to the sun from freezing point to boiling point.

This is a result of a most instructive character. The amount of heat that would be required to raise a pound of water from freezing point to boiling point would, speaking generally, be quite enough if applied to a pound of stone or iron to raise either of these masses to a red heat. If, therefore, we think of the sun as a mighty globe of stone or iron, the amount of heat that would be produced by the contraction of the sun from the primæval nebula would suffice to raise that globe of stone or iron from freezing point up to a red heat 270,000 times. This will give us some idea of the stupendous amount of heat which has been placed at the disposal of the solar system by the process of contraction of the sun. This contraction is still going on, and consequently the yield of heat which is the consequence of this contraction is still in progress, and the heat given out provides the annual supply necessary for the sustenance of our solar system.

There is one point which should be specially mentioned in connection with this argument. We have here supposed that the current supply of radiant heat from the sun is entirely in virtue of the sun's contraction. That is to say, we suppose the sun's temperature to be remaining unaltered. This is perhaps not strictly the case. There may be reason for believing that the temperature of the sun is increasing, though not to an appreciable extent.

It will be convenient to introduce a unit that will be on a scale adapted to our measurements. Let us think of a globe of coal as heavy as the sun. Now suppose adequate oxygen were supplied to burn that coal, a definite quantity of heat would be produced. There is no present necessity to evaluate this in the lesser units adapted for other purposes. In discussing the heat of the sun, we may use what we call the *coal-unit*, by which is to be understood the total quantity of heat that would be produced if a mass of coal equal to the sun in weight were burned in oxygen. It can be shown by calculations, which will be found in the Appendix, that in the shrinkage of the sun from an infinitely great extension through space down to its present bulk the contraction would develop the stupendous quantity of heat represented by 3,400 coal-units. It is also shown that one coal unit would be adequate to supply the sun's radiation at its present rate for 2,800 years.

Chapter 7

THE HISTORY OF THE SUN

We pointed out in the last chapter how, in consequence of its perennial loss of heat, the orb of day must be undergoing a gradual diminution in size. In the present chapter we are to set down the remarkable conclusions with respect to the early history of the sun to which we have been conducted by pursuing to its legitimate consequences the shrinkage which the sun had undergone in times past.

The outer circle in Figure 20 represents the track in which our earth now revolves around the sun, and we are to understand that the radius of this circle is about ninety-three million miles. We must imagine that the innermost of the four circles represents the position of the sun. Along its track the earth revolves year after year; so it has revolved for centuries, so it has revolved since the days of the first monarch that ever held sway in Britain, so it has revolved during all the time over which history extends, so it has doubtless revolved for illimitable periods anterior to history. For an interval of time that no one presumes to define with any accuracy the earth has revolved in the same track round that sun in heaven which, during all those ages, has dispensed its benefits of light and heat for the sustenance of life on our globe.

The sun appears constant during those few years in which man is allowed to strut his little hour. The size of the sun and the lustre of the sun has not appreciably altered. But the sun does not always remain the same. It has not always shone with the brightness and vigour with which it shines now; it will not continue forever to dispense its benefits with the same liberality that it does at present.

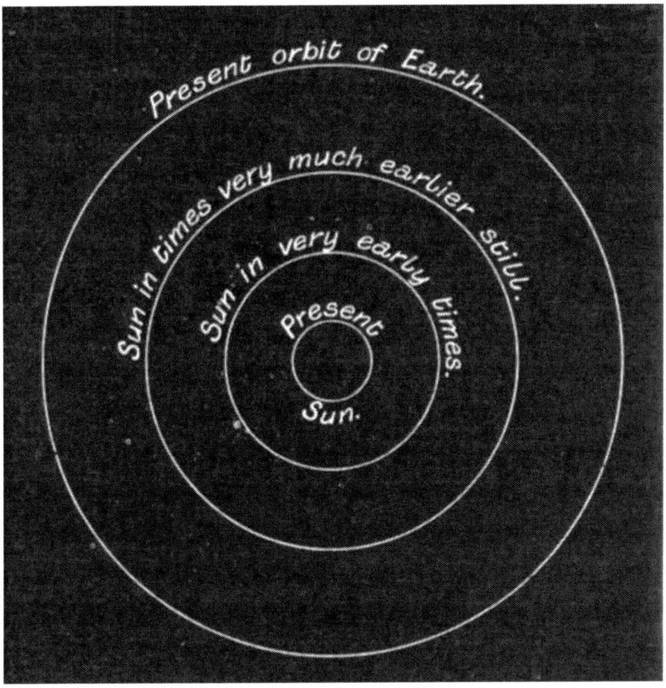

Figure 20. To Illustrate the History of the Sun. Present orbit of Earth. Sun in times very much earlier still. Sun in very early times. Present Sun.

The sun is always in a state of change. It would not indeed be correct to refer to these changes as growths, in the same sense in which we speak of the growth in a tree. Decade after decade the tree waxes greater; but the sun, as we have already explained, does not increase with the time, for the change indeed lies the other way. It may well be that in this present era the sun is near its prime, in so far as its capacity to radiate warmth and brightness is concerned. It is, however, certain that the sun is not now so large as it was in ancient days. The diminution of the orb is still in progress. In these present days of its glorious splendour the orb of day is much larger than it will be in that gloomy old age which destiny assigns to it.

We have already shown how to give numerical precision to our facts. We have stated that the sun's diameter is diminishing at the rate of one mile every eleven years. We have dwelt upon the remarkable significance of that shrinkage in accounting for the sustentation of the sun's heat. We have now to call on this perennial diminution of the sun's diameter to provide some information as to the early history of our luminary.

The innermost circle in our sketch is to suggest the sun as it is at present. Millions of years ago the orb of day was as large as I have indicated it by the circle with the words "sun in very early times." It will, of course, be understood that we do not make any claim to precise representation of the magnitude of the orb. At a period much earlier still, the sun must have been larger still, and we venture so to depict it. We know the rate at which the sun is now contracting, and doubtless this rate has continued sensibly unaltered during thousands of years, and indeed we might say scores of thousands of years. But it would not be at all safe to assume that the annual rate of change in the sun's radius has remained the same throughout excessively remote periods in its evolutionary history. What we do affirm is, that in the course of its evolution the sun must have been contracting continually, and we have been able to learn the particular rate of contraction characteristic of the present time. But though we are ignorant of the rate of contraction at very early epochs, yet the sun ever looms larger and larger in days earlier and still earlier. But in those early days the sun was not heavier, was not, indeed, quite so heavy as it is at present. For we remember that the sun is perennially adding thousands of tons to its bulk by the influx of meteors. Perhaps we ought to add that the gain of mass from the meteors may be to some extent compensated by the loss of substance which the sun not infrequently experiences if, as is sometimes supposed, it expels in some violent convulsion a mass of material which takes the form of a comet.

Let us now consider what the density of the sun must have been in those primæval days, say, for example, when the luminary had ten times the volume that it has at present. Even now, as already stated, it does not weigh half as much again as a globe of water of the same size, so that when it was ten times as big its density must have been only a small fraction of that of water. But we may take a stage still earlier. Let us think of a time—it was, perhaps, many scores of millions of years ago—when the sun was a thousand times as big as it is at present. The same quantity of matter which now constitutes the sun was then expanded over a volume a thousand times greater. A remarkable conclusion follows from this consideration. The air that we breathe has a density which is about the seven-hundredth part of that of water. Hence we see that at the time when the materials of the sun were expanded into a volume a thousand times as great as it is at present the density of the luminary must have been about equal to that of ordinary air. We refer, of course, in such statements to the average density of the sun. It will be remembered that the density of the sun cannot be uniform. The mutual attractions and pressures of the particles in the interior must make the density greater the nearer we approach to the centre.

We must push our argument further still. We have ascertained that the primæval sun could not have been a dense solid body like a ball of metal. It must have been more nearly represented by a ball of gas. There was a time when that collection of matter which now constitutes the sun was so big that a balloon of equal size, filled at ordinary pressure with the lightest of known gases, would contain within it a heavier weight than the sun. At this early period the sun must have been as light as an equal volume of hydrogen. The reasoning which has conducted us to this point remains still unimpaired. From that early period we may therefore look back to periods earlier still. We see that the sun must have been ever larger and larger, for the same quantity of material must have been ever more and more diffused. There was a time when the mean density of the sun must have been far less than that of the gas in any balloon.

We must not pause to consider intermediate stages. We shall look back at once to an excessively early period when the sun—or perhaps we ought rather to say the matter which in a more condensed form now constitutes the sun—was expanded throughout the volume of a globe whose radius was as great as the present distance from the sun to the earth. Have we not here truly an astonishing result, deduced as a necessary consequence from the fundamental laws of heat?

Figure 21. The Solar Corona (January 1st, 1899).
(Photographed during Eclipse by Professor W. H. Pickering.)

I need hardly say that the sun at that early date did not at all resemble the glorious orb to which we owe our very existence. The primæval sun must have been a totally different object, as we can easily imagine if we try to think that the sun's materials then filled a volume twelve million times as great as they occupy at present. Instead of comparing such an object with the gases in our ordinary atmosphere, it should rather be likened to the residue left in an exhausted receiver after the resources of chemistry have been taxed to make as near an approach as possible to a perfect vacuum.

We can give a familiar illustration of gas in a state of extreme tenuity. Look at the beautiful incandescent light with which in these days our buildings are illuminated. How brilliantly those little globes shine! The globe has to be most carefully sealed against the outside air. If there were the smallest opportunity for access, the air from outside would rush in and the lamp would be destroyed. In the preparation of such a lamp elaborate precautions have to be taken to secure that the exhaustion of the air from the little globe shall be as nearly perfect as possible. Of course it is impossible to remove all the air. No known processes can produce a perfect vacuum. Some traces of gas would remain after the air-pump had been applied even for hours.

We must now imagine a globe, not merely two inches in diameter like one of these little lamps, but a globe 186,000,000 miles in diameter, a globe so large that the earth's orbit would just form a girdle round it. Even if this globe had been exhausted, so that its density was only the twelve-thousandth part of the ordinary atmospheric density, it would still contain more material than is found in the sun in heaven. Thus our reasoning has conducted us to the notion of an epoch when the sun—or rather I should say the matter composing the sun—formed something totally different from the orb which we know so well. The matter in that very diffuse state would not dispense light and heat as a sun in the sense in which we understand the word. However vast might be the store of energy which it contained—a store indeed thousands of times greater than our present sun possesses—yet it would hardly be possessed of the power of effective radiation. It would assuredly not be able to warm and light a world associated with it, in the same way as the sun now provides so gloriously for our wants and comfort.

But it is certain that in those early days there was no earth to be warmed and lighted. Our globe, even if it can be said to have existed at all, was truly "without form and void." At the time when the sun was swollen into a great globe of gas or rarefied matter, the elementary substances which were to form the future earth were in a condition utterly different from that of our present globe.

Figure 22. The Great Comet of 1882.
(Photographed on November 7th, 1882, by Sir David Gill, K.C.B.)

The history of this earth itself involves another chapter of the argument. Let it suffice to notice, for the present, that our reasoning has led us to a time when the sun consisted only of a rarefied gaseous material, and let us give to the matter in this condition the name which astronomers apply to any object of a similar character wherever they may meet with it in the universe. Suppose that we could observe through our telescopes at the present moment an object in remote space which was like what the sun must have been at that early stage of its existence which we have been considering, I do not think that the object would be unfamiliar to astronomers. There is, indeed, no doubt that there are many objects visible at this moment, and nightly studied in our observatories, which are formed of matter just in the same state as the sun was in those early times. Examined with a good telescope, the object would seem like a small stain of light on the black background of the sky. The observer would at once call it a nebula. In these modern days he would probably apply the spectroscope to it, and this instrument would assure him that the object he was looking at was a mass of incandescent gas. Such an object would in all probability not greatly differ from many nebulæ now known to us.

This being so, why should we withhold from the sun of primitive days the designation to which it seems to be so fully entitled? Why should we not speak of it as a nebula? The application of the laws of heat has shown that the great orb of day was once one of those numerous objects which astronomers know as nebulæ, and perhaps it may not be too fanciful to suppose that a trace of the primæval nebula still survives in what we call the Solar Corona.

Chapter 8

THE EARTH'S BEGINNING

In the last chapter we endeavoured to ascertain what can be learned from the radiation of the sun with regard to the history of the solar system. In this chapter we shall not consider anybody in the heavens, but the condition of the earth itself. We have learned something of the history of the solar system from the celestial bodies; we shall now learn something about it in another way—from the condition of our globe at depths far beneath our feet.

It will be convenient to commence by mentioning a remarkable experiment which was made a few years ago. Though that experiment is of great scientific interest, yet it was not designed with any scientific object in view. Not less than £10,000 was expended on the enterprise, and probably so large a sum has never been expended on a single experiment of which the sole object was to add to scientific knowledge. In the present case the immediate object in view was, of course, a commercial one. There was, it may be presumed, reasonable expectation that the great initial cost, and a handsome profit as well, would be returned as the fruits of the enterprise. Whether the great experiment was successful from the money-making point of view does not now concern us, but it does concern us to know that the experiment was very successful in the sense that it incidentally afforded scientific information of the very highest value.

The experiment in question was made in Germany, at Schladebach, about fifteen miles from Leipzig. It was undertaken in making a search for coal. Some enterprising capitalists consulted the geologists as to whether coal-seams were likely to be found in this locality. They were assured that coal was there, though it must certainly be a very long way down, and consequently the pit by which alone the seams could be worked would have to be unusually deep. The capitalists were not daunted by this consideration. But, before incurring the great expense of sinking the shaft, they determined to make a preliminary search and verily the actual presence of workable seams of useful fuel. They determined to bore a hole down through the rocks deep enough to reach the coal, if it could be reached. A boring for coal was, of course, by no means a novelty; but there was an unprecedented degree of mechanical skill and scientific acumen shown in this memorable boring near Leipzig. The result of this enterprise was to make the deepest hole which, with perhaps a single more recent exception not of so much scientific interest, has ever been pierced through the crust of the earth. This boring was merely a preliminary to the operations which would follow if the experiment were successful in discovering coal. It was accordingly only necessary to make a hole large enough to allow specimens of the strata to be brought to the surface.

The instrument employed in sinking a hole of such a phenomenal depth through solid rock is characteristic of modern enterprise. The boring tool had a cutting edge of diamonds: for no other cutting implement is at once hard enough and durable enough to advance steadily, yard by yard, through the various rocks and minerals that are met with in the descent through the earth's crust. We might, perhaps, illustrate the actual form of the tool as follows: imagine a piece of iron pipe, about six inches in diameter, cut squarely across, with diamonds inserted round its circular end, and we have a notion of the diamond drill. If the drill be made to revolve when held vertically, with the diamonds in contact with the rocks, the cutting will commence. As the rotation is continued, the drill advances through the rocks, and a solid core of the material will occupy the hollow of the pipe. We do not now enter into any description of the many mechanical details; there are ingenious contrivances for removing the *débris* produced by the attrition of the rocks as the diamonds cut their way, and provision is also made for carefully raising the valuable core which, as it provides specimens of the different strata pierced, will show the coal, if coal is ever reached. There is, of course, an arrangement by which, as the first length of drill becomes buried, successive lengths can be added, so as to transmit the motion to the cutting edge and enable the tool to be raised when necessary; in this manner one length of solid rock after another is brought up for examination. These cores, when ranged in series, give to the miner the information he requires as to the different beds of rock through which the instrument has pierced in its descent and as to the depths of the beds. A series of cores will sometimes show astonishing variety in the material through which the drill has passed. Here the tool will be seen passing through a bed of hard limestone, and then entering a bed of soft shale; now the tool bores through dense and hard masses of greenstone, anon it pierces, it may be, a stratum of white marble; and finally the explorer may hope to find his expectations realised by the arrival at the surface of a cylinder of solid coal.

The famous boring to which we are now referring, though very deep, was not large in diameter. As it descended the comparatively large tool first employed was replaced by a succession of smaller tools, so that the hole gradually tapered from the surface to the lowest point. At its greatest depth the hole was indeed hardly larger than a man's little finger. It increased gradually all the way to the surface, where it was large enough for a man's arm to enter it easily.

How often do we find that the success which rewards mechanical enterprise greatly transcends even the most sanguine estimate previously formed! Without the actual experience which has been acquired, I do not think anyone could have anticipated the extraordinary facilities which the diamond drill has given in the operations of a deep boring. This hole at Schladebach was, indeed, a wonderful success. It pierced deeper than any previous excavation, deeper than any well, deeper than any coal pit. From the surface of the ground, where the hole was some six inches in diameter, down to the lowest point, where it was only as large as a little finger, the vertical depth was not less than one mile and a hundred and seventeen yards.

It is worth pondering for a moment on the extraordinary mechanical feat which this represents. When the greatest depth was reached, the total length of the series of boring rods from the surface where the machinery was engaged in rotating the tool down to the cutting diamonds at the lower end where the penetration was being effected, was as long as from Piccadilly Circus to the top of Portland Place. If a hole of equal length had been bored downwards from the top of Ben Nevis, it would have reached the sea level and gone down 1,200 feet lower still. When the foreman in charge wished to look at the tool to see whether it was working satisfactorily, or whether any of the diamonds had got injured or displaced, it was necessary to raise that tremendous series of rods. Each one of them had to be lifted, had to be uncoupled, and had to be laid aside. I need hardly say that such an operation was a very tedious one. The collective weight of the working system of rods was about twenty tons, and not less than ten hours' hard work was required before the tool was raised from the bottom to the surface. We may, I believe, conclude that so much ingenuity and so much trouble was never before expended on the act of boring a hole; but the results are full of information on important problems of science.

I am not going to speak of the geological results of this exploration. There is not the least doubt that the remarkable section of the earth's crust thus obtained is of much interest to geologists. Our object in now alluding to this wonderful boring is, however, very different. Its significance will be realised when we say that it gives us more full and definite information about the internal heat of the earth than had ever been obtained by any other experiment on the earth's crust. No doubt many previous observations of the internal heat of the globe were well known to the investigators who feel an interest in these important questions; but the exceptional depth of this boring, as well as the exceptionally favourable conditions under which it was made, have rendered the information derived from it of the utmost value to science.

We ought first to record our special obligation to the German engineer, Captain Huyssen, who bored this wonderful hole. He was not only a highly skilful mining engineer, diligent in the pursuit of his profession, but, by the valuable scientific work he has done, he has shown himself to be one of those cultivated and thoughtful students who love to avail themselves of every opportunity of searching into Nature's secrets. Our thanks are due to him for the remarkable zeal with which he utilised the exceptional opportunities for valuable scientific work that arose, incidentally as it were, in connection with the work committed to him.

Of course, everybody knows that the temperature of the earth is found to increase gradually as greater depths are reached. The rate at which the increase takes place has been determined on many occasions. But when opportunities have arisen for taking the temperature at considerable depths below the earth's surface, it has happened sometimes that the observations have been complicated by circumstances which deprived them of a good deal of their accuracy. If our object be to learn the law connecting the earth's temperature with the depth below the surface, it is not sufficient to study the thermometric readings in different coal pits. Throughout the workings in every pit there must be arrangements for ventilation. The cool air has to be drawn down, and thus the temperature indicated in the pit is forced below the temperature which would really be found at that depth if external sources of change of temperature were absent.

Captain Huyssen rightly deemed that the hole which he had pierced presented exceptional opportunities for the study of the important question of the earth's internal temperature. Precautions had, of course, to be observed. The hole, as might be expected, was filled with water, and the water would tend, if its circulation were permitted, to equalise the temperature at different depths. But the ingenious Captain quickly found an efficient remedy for this source of inaccuracy. He devised an arrangement, which I must not delay to describe, by which he could place temporary plugs in the hole at any depths he might desire; he then determined the temperature of the water in a short length, so plugged above and below that the circulation was stopped, and accordingly the water thus confined might be relied on to indicate the temperatures of the strata which hold it.

The thermometer employed in an investigation of this sort is ingenious though extremely simple. The ordinary maximum thermometer is not found to be adapted for the purpose. The instrument employed in the determination of underground temperatures is very much less complicated and at the same time much more accurate. The contrivance is indeed so worthy of notice that I do not like to pass it by without a few words. The thermometer with which the temperature of the earth is ascertained in such investigations is not like any ordinary thermometer. There is no scale of degrees attached to it or engraved upon it, as we generally find in such instruments. The instrument with which the temperature of the deep hole was measured was merely a bulb of glass with a slender capillary stem, the end of which was not closed. When it was about to be lowered to test the temperature of the rocks at the lowest point to which the drill had penetrated, the bulb and the tube were first filled with mercury to the top, and brimming over. This simple apparatus was attached to a long wire, by the aid of which it could be lowered down this deep hole. Down it went till at last the thermometer reached the bottom, which, as we have explained, it could not do until more than a mile of wire had been paid out. The instrument was then left quietly until it presently assumed the same temperature as the rocks about it. There could be no interference by heat from other strata, as the circulation of water was prevented by the plugging already referred to. The temperature to which the thermometer had been exposed must, therefore, have been precisely the temperature corresponding in that particular locality to that particular depth below the earth's surface.

As the thermometer descended, it passed through a succession of strata of ever-increasing temperature. Consequently the mercury, which, it will be remembered, had completely filled the instrument when it was at the surface, began to expand according as it was exposed to greater temperatures.

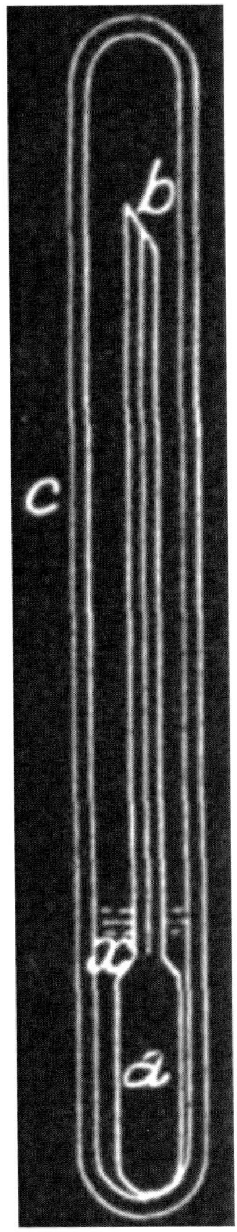

Figure 23. Special Thermometer for Use in Deep Borings.

As the mercury expanded, it must, of course, flow out of the tube and be lost, because the tube had been already full. So long as the mercury was gaining in temperature, more and more of it escaped from the top of the tube, and the flow only ceased when the instrument was resting at the bottom of the hole, and the mercury became as hot as the surrounding rocks. No more mercury was then expelled, the tube, however, remaining full to the brim. After allowing a sufficient time for the temperature to settle definitely, the thermometer was raised to the surface. As it ascended through the long bore the temperature surrounding it steadily declined. With the fall in the temperature of the mercury the volume of that liquid began to shrink; but the mercury already expelled could not be recalled. When at last the instrument had safely reached the surface, after its long journey down and up, and when the mercury had regained the temperature of the air, the lessened quantity that remained told the tale of the changes of temperature.

It is now easy to see how, even in the absence of an engraved scale on the instrument, it is possible to determine, from the amount of mercury remaining, the temperature to which the thermometer has been subjected at the bottom of the boring. It is only necessary to place this thermometer in a basin of cold water, and then gradually increase the temperature by adding hot water. As the temperature increases the mercury will, of course, rise, and the hotter the water the more nearly will the mercury approach the top of the tube. At last, when the mercury has just reached the top of the tube, and when it is just on the point of overflowing, we may feel certain that the temperature of the water in the basin has been raised to the same temperature as that to which the instrument was subjected at the bottom of the boring. In each case the temperature is just sufficient to expand the quantity of mercury remaining in the instrument so as to make it fill precisely both bulb and stem. When this critical condition is reached, it only remains to dip a standard thermometer, furnished with the ordinary graduation, into the hot water of the basin. Thus we learn the temperature of the basin, thus we learn the temperature of the mercury in the thermometer, and thus we determine the temperature at the bottom of the boring over a mile deep.

I need not specify the details of the arrangements which enabled the skilful engineer also to determine the temperature at various points of the hole intermediate between the top and the bottom. In fact, taking every precaution to secure accuracy, he made measurements of the temperature at a succession of points about a hundred feet distant throughout the whole depth. In each case he was careful, as I have already indicated, to plug the hole above and below the thermometer, so as to prevent the circulation of water in the vicinity of the instrument. The thermometer, therefore, recorded the temperature of the surrounding rocks without any disturbing element. Fifty-eight measurements at equal distances from the surface to the greatest depths were thus obtained.

We have now to discuss the instructive results to which we have been conducted by this remarkable series of measurements. First let us notice that there is much less variation in the subterranean temperatures than in the temperatures on the earth's surface. On the surface of the earth we are accustomed to large fluctuations of temperature. We have, of course, the diurnal fluctuations in temperature from day to night; we have also the great seasonal fluctuations between summer and winter. But below a certain depth in the ground the temperature becomes much more equable. Whether the temperature on the surface be high or whether it be low, the temperature of any particular point far beneath the surface does not change to any appreciable extent. In Arctic regions the surface of the earth may undergo violent seasonal changes of temperature, while at a few feet below the surface the temperature, from one end of the year to the other, may remain sensibly unaltered.

In deep and extensive caverns the temperature is sometimes found to remain practically unaffected by the changes in the seasons. The Mammoth Cave of Kentucky is a notable instance. The uniformity of the temperature, as well as the mildness and dryness of the air, in those wonderful subterranean vaults is such that many years ago a project was formed to utilise the cavern as an abode for consumptive patients, for whose cure, according to the belief then prevailing, an equable temperature was above all things to be desired. Houses were indeed actually built on the sandy floors of the cavern, and I believe they were for some time tenanted by consumptive patients willing to try this desperate remedy. The temperature may have been uniform and the air may have been dry, but the intolerable gloom of such a residence entirely neutralised any beneficial effects that might otherwise have accrued. The ruins of the houses still remain to testify to the failure of the experiment.

The heat received from the sun does not penetrate far into the earth's crust, and consequently the diurnal and even the seasonal changes of the temperature at the surface produce less and less effect with every increase of the depth. All such variations of temperature are confined to within 100 feet of the surface. At the depth of about 100 feet a fixed temperature of 52° Fahrenheit is reached, and this is true all over the earth. It matters not whether the surface be hot or cold, whether the latitude is tropical and the season is midsummer, whether the latitude lie in the Arctic regions and the season be the awful winter of iron-bound frost and total absence of sun—in all cases we find that about 100 feet below the surface the temperature is 52°. With sufficient accuracy we may say that this depth expresses the limit of the penetration of the earth's crust by sunbeams. The remarkable law according to which the temperature changes below the depth of 100 feet is wholly irrespective of the solar radiation.

The study of the internal heat of the earth may be said to begin below the level of 100 feet, and the results that were obtained in the great boring are extremely accordant. The deeper the hole, the hotter the rocks; and Captain Huyssen found that for each sixty-six feet in descent the temperature increased one degree Fahrenheit. To illustrate the actual observations, let us take two particular cases. We have said that the hole was one mile and 117 yards deep. Let us first suppose the thermometer to be lowered 117 yards and then raised, after a due observance of the precautions required to obtain an accurate result. The temperature of the rocks at the depth of 117 yards is thus ascertained. In the next observation let the thermometer be lowered from the surface to the bottom of the hole, that is to say, exactly one mile below the position which it occupied in the former experiment. The observations indicate a temperature 80° Fahrenheit higher in the latter case than in the former. We have thus ascertained a most important fact. We have shown that the temperature of the crust of the earth at the depth of one mile increases about 80°. This is at the rate of one degree every sixty-six feet. I should just add, as a caution, that if we choose to say the temperature increases one degree per sixty-six feet of descent, we ought to suppose that we start from a point which is not higher than that level of 100 feet above which as already explained, the temperature of the rocks is more or less affected by solar heat.

We have described these particular observations in some detail because they have been conducted under conditions far more favourable to accuracy than have ever been available in any previous investigations of the same kind. But now we shall omit further reference to this particular undertaking near Leipzig. It is not alone in that particular locality, not alone in Germany, not alone in Europe, not alone on the surface of any continent, that this statement may be made. The statement is one universally true so far as our whole earth is concerned. Wherever we bore a hole through the earth's crust, whether that hole be made in the desert of Sahara or through the icebound coasts of Greenland, we should find the general rule to obtain, that there is an increase of temperature of about 80° for a mile of descent. This is true in every continent, it is true in every island; and, though we cannot here go into the evidence fully, there is not the least doubt that it is true also under the floor of ocean. If beneath the bed of the Atlantic a hole a mile deep were pierced, the temperature of the rocks at the bottom of that hole would, it is believed, exceed by about 80° the temperature of the rocks at the surface where the hole had its origin. We learn that at the depth of a mile the temperature of the earth must generally be 80° hotter than it is at the level of constant temperature near the surface.

It may perhaps help us to realise the significance of this statement if we think of the following illustration. Let us imagine that the waters of the ocean were removed from the earth. The ocean may in places be five or six miles deep, but that is quite an inconsiderable quantity when compared with the diameter of the earth. The change in the size of the earth by the removal of all the water would not be greater, proportionally, than the change produced in a wet football by simply wiping it dry. Let us suppose that an outer layer of the earth's surface, a mile in thickness, was then to be peeled off. If we remember that the diameter of the earth is 8,000 miles, we shall see that this outer layer, whose removal we have supposed, does not bear to the whole extent of the earth a ratio even as great as that which the skin of a peach does to the fruit inside. But this much is certain, that if the earth were so peeled there would be a wonderful difference in its nature. For though practically of the same size as it is at present, it would be so hot that it would be impossible to live upon it.

Next comes the very interesting question as to the temperature that would be found at the bottom of a hole deeper still than that we have been considering. Our curiosity as to the depths of the earth extends much below the point to which Captain Huyssen drove down his diamond drill. The trouble and the cost of still deeper exploration of the same kind seem, however, to be actually prohibitive. To bore a hole two miles deep would certainly cost a great deal more than twice the sum which sufficed to bore a hole one mile deep. At a great depth each further foot could only be won with not less difficulty and expense than a dozen, or many dozen feet, at the surface. Mining enterprise does not at present seem to contemplate actual workings at depths much over a mile, so there does not seem much chance of any very much deeper boring being attempted. We do not say that a hole two miles deep would be actually impossible; it may well be wished that some millionaire could be induced to try the experiment. We should greatly like to be able to lower a thermometer down to a depth of two miles through the earth's crust.

Seeing there is but little chance of our wish for such future experiments being gratified, it is consolatory to find that actual observations of this kind are not indispensable to the argument on which we are to enter. Our argument can indeed be conducted a stage further, even with our present information. The indications already obtained in the hole one mile deep go a long way towards proving what the temperature of a hole still deeper would be. We have already remarked that it was part of Captain Huyssen's scheme to obtain careful readings of his thermometer at intervals of 100 feet from the surface to the bottom of the hole. A study of these readings shows that the increase of 80° in a mile takes place uniformly at the rate of one degree for each sixty-six feet of depth. As the temperature increases uniformly from the surface down to the lowest point which our thermometers have reached, it would be unreasonable to suppose that the rate of increase would be found to suffer some abrupt change if it were possible to go a little deeper. As the temperature rises 80°

in the first mile, and as the rate of increase is shown by the observations to be quite as large at the bottom of the hole as it is at the top, we certainly shall not make any very great mistake if we venture to assume that in the second mile the temperature would also increase to an extent which will not be far from 80°. This inference from the observations leads to the remarkable conclusion that at a depth of two miles the temperature of the earth must be, we will not say exactly, but at all events not very far from, 160° higher than at the level of constant temperature about 100 feet down.

As in the former case, we need not confine ourselves to any particular locality in drawing this conclusion. The arguments apply not only to the rocks underneath Leipzig, but to the rocks over every part of the globe, whether on continents or islands, or even if forming the base of an ocean. No one denies that the law of increase in temperature with the depth must submit to some variation in accordance with local circumstances. In essential features it may, however, be conceded that the law is the same over all the earth. If we take 52° to be the temperature of the level 100 feet down, which limits the seasonal variations, and if we add that at two miles further down the temperature is somewhere about 160° more, we come to the conclusion that at a depth of a little over two miles the temperature of the rocks forming the earth's crust is about 212° Fahrenheit. Thus we draw the important inference that if, the oceans having been removed, we were then to remove from the earth's surface a rind two miles thick—a thickness which, it is to be observed, is only the two-thousandth part of the earth's radius—we should transform the earth into a globe which, while it still retained appreciably the same size, would have such a temperature that even the coolest spot would be as hot as boiling water. This is indeed a remarkable result.

And now that we have gone so far, it is impossible for us to resist making a further attempt to determine what the temperature of the earth's crust must be if we could send a thermometer still lower. A hole one mile deep we have seen; I do not think we can hope to see a hole two miles deep, but still it may not be absolutely impracticable; but a hole of three or more miles deep we may safely regard as transcending present possibilities in engineering enterprise. Are we therefore to be deprived of all information as to the condition of our earth at depths exceeding those already considered? Fortunately we can learn something. We are assisted by certain laws of heat, and, though the evidence on which we believe those laws is necessarily limited to the experience of Nature as it comes within our observation, yet it is impossible to refuse assent to the belief that the same laws will regulate the transmission of heat in the crust of the earth two miles, three miles, or many miles beneath our feet.

I represent, in the diagram shown three consecutive beds of rock—A, B, and C—as they lie in the earth's crust, a little more than a mile beneath our feet. I shall suppose that the bed B is the very lowest rock whose temperature was determined in the great boring. The drill has passed completely through A, it has pierced to the middle of B, but it has not entered C. The observations have shown that the temperature of the stratum B exceeds that of the stratum A, and we further note that this is a permanent condition—that is to say, B constantly remains hotter than A. From this fact alone we can learn something as regards the temperature of the stratum C which lies in contact with B. Of course we are unable to observe the temperature of C directly, because by hypothesis the boring tool has not entered that rock. We can, however, prove, from the laws of the conduction of heat, that the temperature of C must be greater than that of B; and this appears from the following consideration.

It is plain that C must be either just the same temperature as B, or it must be hotter than B, or it must be colder than B. If C were the same temperature as B, then the law of conduction of heat tells us that no heat would flow from one of these strata to the other. The laws of heat, however, assure us that when two bodies at different temperatures are in contact the heat will flow from the hotter of these bodies into the colder, so long as the inequality of temperature is maintained. As B is hotter than A, then heat must necessarily flow from B into A, and this flow must tend to equalise the temperature in these strata, for B is losing heat while none is flowing into it from C. Therefore B and A could not continue to preserve indefinitely the different temperatures which observation shows them to do. We are therefore forced to the conclusion that B and C cannot be at the same temperature.

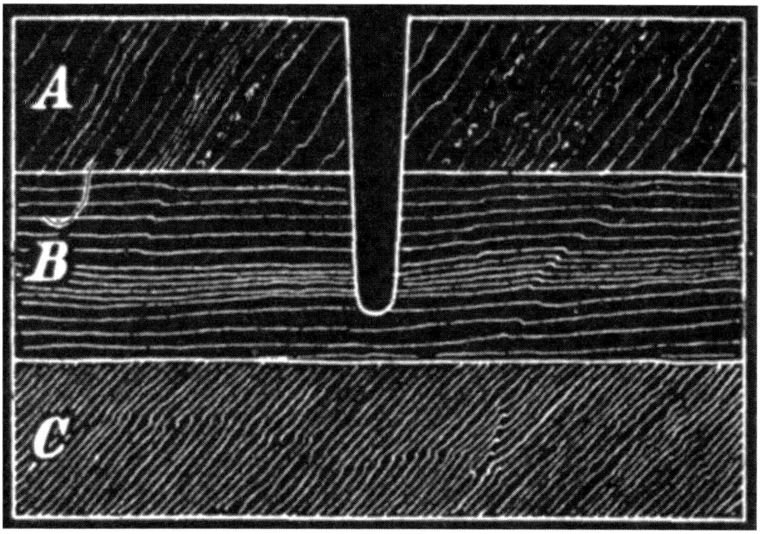

Figure 24. At the Bottom of the Great Bore.

Next let us suppose that the temperature of the stratum B exceeded that of C. Then, as A is colder than B, it appears that B would be lying between two strata each having a temperature lower than itself. But that, of course, cannot be a permanent arrangement, for the heat would then escape from B on both sides. The laws of heat, therefore, tell us that B could not possibly retain permanently a temperature above both A and C. Observation, however, shows that the temperatures of A and B are persistently unequal. We are therefore obliged to reject the supposition that the temperature of C can be less than that of B.

We have thus demonstrated that the temperature of the stratum C cannot be the same as that of B. We have also demonstrated that it cannot be colder than B. We must therefore believe that C is hotter than B. This proves that the stratum immediately beneath that stratum to which the observations have extended must be hotter than it. Thus, though the stratum below the bottom of the hole lies beyond the reach of our actual observation, we have, nevertheless, been able to learn something with regard to its temperature.

Having established this much, we can continue the same argument further; indeed, it would seem that we can continue it indefinitely, so long as there is a succession of such strata. Underneath the stratum C must lie another stratum D. But we have shown that C must be hotter than B, and precisely the same argument that has proved this will prove that D is hotter than C. Underneath D comes the stratum E, and again the same argument will apply. Inasmuch as D is hotter than C, it follows that E must be hotter than D. These three strata, C, D, and E, are all beyond the reach of the thermometer, we know nothing of their temperatures by direct observation; but none the less is the argument, which we are following strictly, applicable. Thus we obtain the important result that in the crust of the earth the temperature must be always greater, the greater the depth beneath the surface.

We have seen that the rate of increase of temperature with the depth is about 80° for the first mile, and we deem it probable that the rate of increase may be maintained at about the same for the second mile. But we do not suppose that the rate of increase mile after mile will remain the same at extremely great depths. It may perhaps be presumed that there must be some increase of temperature all the way to the earth's centre; but the rate of increase per mile may change as the centre is approached. The point of importance for our present argument is, that the temperature of the earth must increase with the depth, though the rate of increase is quite unknown to us at depths greatly beyond those which the thermometer has reached. It is easy to see that the conditions prevailing in the earth's interior might greatly modify any conclusion we should draw from observations near the surface. Our argument has been based on the laws of heat, as we find them existing in matter on the surface of the earth submitted to such ranges of different physical conditions as can be dealt with in our laboratories; but at such excessively high temperatures as may exist in the earth's interior the properties of matter may be widely different from the properties of matter as known to us within the temperatures that we are able to produce and control. The enormous pressure to which matter in the interior of the earth must be subjected should also be mentioned in this connection. It is wholly impossible to produce pressures by any mechanical artifice which even distantly approach in intensity to that awful force to which matter is subjected in the earth's interior.

It may be instructive to consider a few facts with respect to this question of pressure in the Earth's interior. A column of water 34½ feet high gives, as everybody knows, a pressure of fifteen pounds on the square inch. It will be quite accurate enough for our present purpose to assume that the average density of rock is three times that of water: the pressure of ten feet of rock would therefore produce the same pressure as thirty feet of water, that is to say, fifteen pounds on the square inch. The pressure due to the superincumbent weight of a mile of rock would be more than three tons on the square inch. At the depth of ten miles beneath the earth's surface the pressure, amounting as it does to over thirty tons on the square inch, would very nearly equal the pressure produced on the inside of a 100-ton gun when the charge of cordite has been exploded to drive the missile forth. This is indeed about as large a pressure as can well be dealt with artificially, for we know that the 100-ton gun has to be enormously strong if it is to resist this pressure. But ten miles of rock is as nothing compared with the thickness of rock that produces the pressures in the earth's interior. Even if a shell of rocks ten miles thick were removed from the surface it would alter the diameter of our globe by no more than one four-hundredth part. At the depth of about thirty miles from the surface the pressure in the earth's interior would amount to some 100 tons on each square inch. With each increase in depth the pressure increases enormously, though it may not be correct to say that the pressure is proportional to the depth. A pressure of 1,000 tons on the square inch must exist at a depth which is still quite small in comparison with the radius of the earth.

We have not, and apparently cannot have, the least experimental knowledge of the properties of matter at the moment when it is subjected to pressure amounting to thousands of tons per square inch; still less can we determine the behaviour of matter at that pressure of scores of thousands of tons, to which much of the interior of the earth is at this moment subjected. Professor Dewar, in his memorable researches, has revealed to us the remarkable changes exhibited in the properties of matter when that matter has been cooled to a temperature which lies in the vicinity of absolute zero. We can, however, hardly hope that any experiments will give us information as to the properties of matter when heated to a temperature vastly transcending that which could ever be produced in our most powerful electric furnaces, and at the same time exposed to a pressure hundreds of times, or indeed we may say thousands of times, greater than any pressure that has ever been produced artificially by the action of the most violent explosive with which the discoveries of chemistry have made us acquainted.

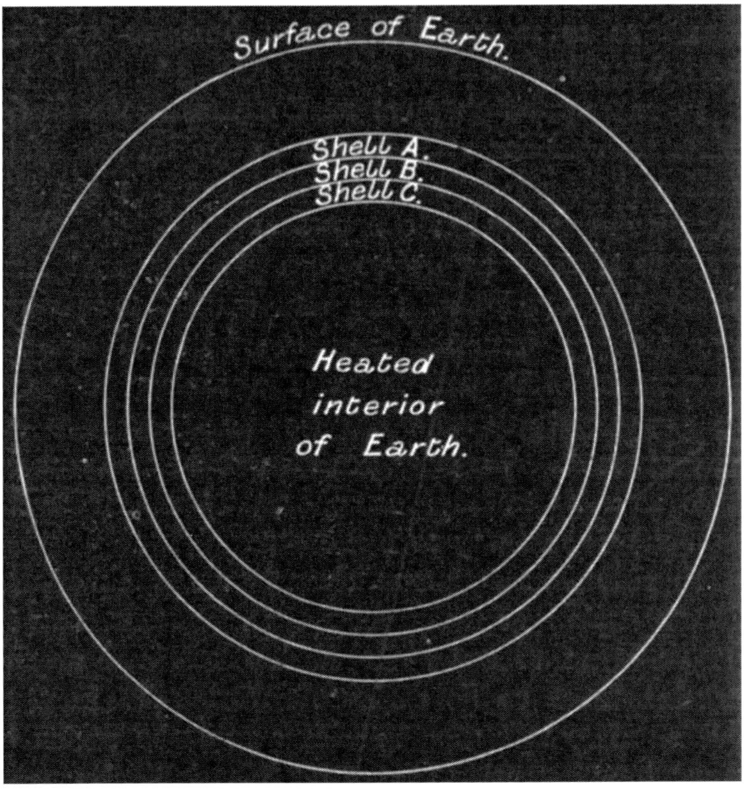

Figure 25. Three Consecutive Shells of the Earth's Crust.

We really do not know how far the laws of heat, which have been employed in showing that the temperature must increase as the depth increases, can be considered as valid under the extreme condition to which matter is subjected in the deep interior of our globe. The laws may be profoundly modified. It suffices, fortunately for our present argument, to say that, so far as observations have been possible, the temperature does gradually increase with the depth, and that this increase must still continue from stratum to stratum as greater depths are reached, unless it should be found that by the excessive exaltation of temperature and the vast intensity of pressure certain properties of matter become so transformed as to render the laws of heat, as we know them, inapplicable.

In subsequent chapters we shall have some further points to consider with respect to the interior of the earth and its physical characteristics, which are, however, not necessary for our present argument. What we now desire to prove can be deduced from the demonstrated fact that the earth's temperature does steadily increase from the level of constant temperature, 100 feet below the surface, down to the greatest depth to which thermometers have ever been lowered. We may presume that the same law holds at very much greater depths, even if it does not hold all the way to the centre.

To make our argument clear, let us think of three different strata of rock. This time, however, we shall suppose them to cover the whole earth, and we shall consider them to lie within the first mile from the surface; they will thus be well within the region explored by observation. We shall also regard them as shells of uniform thickness, and it will be convenient to think of them as being so very thin that we may consider any one of the shells called A to have practically a uniform temperature. The next shell B immediately inside A will have a slightly greater temperature, and be also regarded as uniform, and the shell immediately inside that again will have a temperature greater still. We shall call the innermost of the three shells C, and C is hotter than the next outer shell B, while B is hotter than A. The laws of heat tell us that as B and A are in contact, and that as B is continually hotter than A, then B must be continuously transmitting heat to A. In fact, B appears to be constantly endeavouring to reduce itself to the temperature of A by sharing with A the excess of temperature which it possesses. But if we consider the relation between the shell B and the hotter shell C, immediately beneath it, we see that precisely the same argument will show that B is constantly receiving heat from C. We thus see that while B is continuously discharging heat from its outside surface, it is as constantly receiving heat which enters through its inside surface. Heat enters B from C, and heat passes from B into A, so that B is in fact a channel through which heat passes from C into A.

That which we have shown to take place in those three consecutive layers in the earth's crust must also take place in every three consecutive layers. Each layer is continually receiving heat from the layer below, and is as constantly communicating heat to the layer above. No doubt the rocks are very bad conductors of heat, so that the transmission of heat from layer to layer is a very slow process. But even if this flow of heat be slow, it is incessant, so that in the course of ages large quantities of heat are gradually transmitted from the earth's interior, and ultimately reach the level of constant temperature. There is nothing, however, to impede their outward progress, so at last the heat reaches the earth's surface.

When the surface has been reached, then another law of heat declares what must happen next. It is, of course, by conduction that the heat passes from layer to layer in its outward progress, until it ultimately gains the surface. At the surface the heat is then absolutely removed from the solid earth either by the convection through the air or by direct radiation into space.

I may here interrupt the argument for a moment to make quite clear a point which might perhaps otherwise offer some difficulty to the reader. When this outward flow of heat reaches the superficial layers it becomes, of course, mixed up with the heat which has been absorbed by the soil from the direct radiation of the sun, and this varies, of course, with the hour of the day and with the season of the year. The heat which steadily leaks from the interior has an effect on the rocks near the surface, which is only infinitesimal in comparison with the heat which they receive from periodic causes. We may, however, say that whatever would be the temperature of the rock, so far as the periodic causes are concerned, the actual temperature is always to some minute extent increased by reason of the heat from the earth's interior. The argument is, perhaps, still clearer if, instead of attending to the earth's surface, we think only of that shell, some 100 feet down, which marks the limit of the depth to which the seasonal and diurnal variations of heat extend. The argument shows how the internal heat of the earth, passing from shell to shell in the interior, reaches this layer of constant temperature, and passing through it, enters into those superficial strata of the earth which are exposed to the seasonal variations. With what befalls that heat ultimately we need not now concern ourselves; it suffices for our argument to show that there is a current of heat outward across this level. It is a current which is never reversed, and consequently must produce a never-ceasing drainage from the heat with which it would seem that the interior of the earth is so copiously provided.

Calculations have been made to ascertain how much heat passes annually from the earth's interior, across this surface of constant temperature, out into the superficial regions from which in due course it becomes lost by radiation. A convenient way of measuring a quantity of heat is by the amount of ice it will melt, for of course a definite quantity of heat is required to melt a definite quantity of ice. It has been estimated by Professor J. D. Everett, F.R.S., that the amount of internal heat escaping from our earth each year would be sufficient to melt a shell of ice one-fifth of an inch thick over the whole surface of the globe. We cannot indeed pretend that any determination of the actual loss of heat which our earth experiences could be very precise. Sufficient observations have not yet been obtained, for the operation is so slow that an immense period would have to elapse before the total quantity of heat lost would be sufficient to produce effects large enough to be measured accurately. But now let us hasten to add that, for the argument as to the nebular theory with which we are at present concerned, it is not really material to know the precise rate at which heat is lost. It is absolutely certain that a perennial leakage of heat from the interior of the earth does take place. This fact, and not the amount of that leakage, is the essential point.

And this loss, which is at present going on, has been going on continually. Heat from the earth has been lost this year and last year; it has been lost for hundreds of years and for thousands of years. Not alone during the periods of human history has the earth's heat been declining. Even throughout those periods, those overwhelming periods which geology has revealed to us, has this earth of ours been slowly parting with its heat.

Let us pursue this reflection to its legitimate consequence. Whatever may ultimately become of that heat, it is certain that once radiated into space it is lost forever so far as this globe is concerned. You must not imagine that the warm beams of the sun possess any power of replenishment by which they can restore to the earth the heat which it has been squandering for unlimited ages; we have already explained that the effect of the heat radiated to us from the sun is purely superficial. Even amid the glories of the tropics, even in the burning heat of the desert, the vertical sun produces no appreciable effects at depths greater than this critical limit, which is about 100 feet below the surface. The rigours of an Arctic winter have as little effect in reducing the temperature of the rocks at that depth as the torrid heat at the Equator has in raising it. The effect in each case is nothing.

The argument which we are here employing to deduce the nebulous origin of our earth from the increase of temperature with increase in depth in the earth's crust must be cleared from an objection. It is necessary to explain the matter fully, because it touches on a doctrine of very great interest and importance.

That a rotating body should possess a quantity of energy in virtue of its rotation will be familiar to anyone who has ever turned a grindstone or watched the fly-wheel of an engine. A certain amount of work has to be expended to set the heavy wheel into rotation, and when the machine is called upon to do work it will yield up energy and its motion will undergo a corresponding abatement. The heavy fly-wheel of the machine in a rolling mill contains, in virtue of its motion, enough energy to overcome the tremendous resistance of the materials submitted to it. Once upon a time the earth revolved upon its axis in six hours, instead of in the twenty-four hours which it now requires. At that time the energy of the rotation must have been sixteenfold what it is at present. This consideration shows that fifteen-sixteenths of the energy that the earth originally possessed in its rotation has disappeared, and we want to know what has become of it.

We are here entering upon a matter of some difficulty. It is connected with that remarkable chapter in astronomy which describes the evolution of the earth-moon system. The moon was originally a part of the earth, for in very early times, when the earth was still in a plastic state, a separation would seem to have taken place, by which a small piece broke off to form the moon, which has been gradually revolving in an enlarging orbit until it has attained the position it now occupies. A considerable portion of the energy of the earth's rotation has been applied to the purpose of driving the moon out to its present path, but there is a large remainder which cannot be so accounted for. It is well known that the evolution of the moon has been a remarkable consequence of tidal action. There are tides which sway to and fro in the waters on the earth's surface; there are tides in any molten or viscous matter that the earth may contain, and there are even certain small tidal displacements in the solid material of our globe. Tides of any kind will generate friction, and friction produces heat, and the energy of the earth's rotation, which we have not been able to account for otherwise, has been thus transformed into heat. Throughout the whole interior of the earth heat has been produced by the tidal displacement of its parts. The question therefore arises as to whether the internal heat of the earth may not receive an adequate explanation from this tidal action, which is certainly sufficient as to quantity. It is easy to calculate what the total quantity of this tidal heat may have been. We know the energy which the earth had when it rotated in six hours, and we know that it now retains no more than a sixteenth of that amount. We know also precisely how much was absorbed in the removal of the moon, and the balance can be evaluated in heat. It can be shown, and the fact is a very striking one, that the quantity of heat thus arising would be sufficient to account many times over for the internal heat of the earth. It might therefore be urged plausibly that the internal heat which we actually find has had its origin in this way. And if this were the case the argument which we are using in favour of the nebular origin of the earth, would be, of course, invalidated.

We may state the issue in a slightly different manner, as follows. Heat there is undoubtedly in the earth; that heat might have come from the primæval nebula as we have supposed, and as in actual fact it did come. But apparently it might have come from the tidal friction. Why then are we entitled to reject the latter view, and say that the tidal friction will not explain the internal heat, and why are we compelled to fall back on the only other explanation?

Lord Kelvin suggested a test for deciding to which of these two sources the earth's internal heat was to be attributed. Professor G. H. Darwin applied the test and decided the issue. We have dwelt upon the rate at which the heat increases with the descent, this rate being about one degree every sixty-six feet. Now the distribution of the heat, if it had come from the tidal action, would be quite different from the distribution which would result from the gradual efflux of heat from the centre in the process of cooling. And, speaking quite generally, we may surmise that the heat produced by tidal friction would be distributed rather more towards the exterior of the earth than at its centre. We might therefore reasonably expect that if the internal heat of the earth arose from tidal friction it would be more uniformly distributed throughout the globe, and there would not be so great a contrast between the high temperature of the interior and the lesser temperatures near the surface as there is when the heat distribution is merely the result of cooling. It has been proved that if the internal heat had its origin from the tidal friction, the rate of increase with the depth would be totally different from what it is actually found to be. It would be necessary to go down 2,000 feet to obtain an increase of one degree, instead of only sixty-six feet, as is actually the case.

Hence we conclude that the increasing heat met with in descending through the earth's crust is not accounted for by tidal friction; it has its origin in the other alternative, namely, from the cooling of the primæval nebula. The heat which was undoubtedly produced by the tidal friction has gradually become blended with the heat from the other, and, as we must now say, the principal source. The facts with regard to the rate of increase with depth thus show that, whatever the tides may have done in producing internal heat, there has been another and a still more potent cause in operation. The important conclusion for our present purpose is that our argument may justly proceed without taking account of the effect of tidal friction.

We are led by these considerations to a knowledge of a great transformation in the nature of our globe which must have occurred in the course of ages. We have seen that this earth is gradually losing heat from its interior, and we have seen that this loss of heat is incessant. From the fountains of heat, still so copious, in the interior the supply is gradually dissipating. Now heat is only a form of energy, and energy, like matter, cannot itself be created out of nothing. There can be no creation of heat in our earth without a corresponding expenditure of energy. If, therefore, the earth is radiating heat, then, as there is no known or, indeed, conceivable source of energy by which an equivalent can be restored, it follows that the earth must have less internal heat now than it had at any earlier period. No doubt the process of cooling is excessively slow. The earth has less internal heat at present than it had a hundred years ago, but I do not suppose that even in a thousand years, or perhaps in ten thousand years, there would be any appreciable decline in the quantity of heat, so far as any obvious manifestations of that heat are concerned. It is, however, certain that the earth must have been hotter, even though there are not any observations to which we can appeal to verify the statement; and as our retrospect extends further and still further through the ages we see that the globe must have been ever hotter and ever still hotter. Whatever be the heat contained in our earth now, it must have contained vastly more heat ten million years ago; how otherwise could the daily leakage of heat for all those ten million years have been supplied? It follows that there must have been much more heat somewhere in our earth ten million years ago than there is at present, and the further our retrospect extends the hotter do we find the earth to have been. There was a time when the temperature of the earth's surface must have been warmed not alone by such sunbeams as fell upon it, but by the passage of the heat from the interior.

No matter how early be the period which we consider, we find the same causes to be in operation. There was a time when, owing to the internal heat, the surface of the earth must have been as hot as boiling water. The loss of heat by radiation must then have taken place much more copiously than it does at present. The argument we are pursuing must therefore have applied with even greater force in those early days. There was a time when the materials at the surface of the earth must have been intensely heated, when they must have even been red-hot. There was a time when the earth's surface must have had a temperature like that of the lava as it issues from a volcano. There must have been a time when the surface of the earth was not even solid, when indeed it was a viscid liquid, and earlier still the liquid must have been more and more incandescent. From that brilliant surface heat was vehemently radiated. Each day the globe was hotter than on the succeeding day. There is no break in the argument. We have to think of this glowing globe passing through those phases through which we know that all matter will pass if only we apply to it sufficient heat. The globe assumed the liquid state from that state which is demanded by a temperature still higher, the state in which the matter is actually in the form of vapour. Even the most refractory substances will take the form of vapour at a very high temperature.

Thus we are conducted to a remarkable conception of the condition in which the materials now forming our solid earth must have been in the exceedingly remote past. What is now our earth must once have been a great quantity of heated vapour. It need hardly be said that in that form the volume of the earth was much larger than the volume which the earth has at present, while no doubt the mass of the earth then was even less than the mass of the earth now, by reason of the meteoric matter which has been drawn in by our globe.

But even when our earth was in this inflated state of vapour our argument can be still maintained. Thus we see that the earth, or rather the cloud of vapour which was ultimately to form the earth, is ever growing larger and larger in our retrospect, ever becoming more and more rarefied; and it may well have been that there was a time when the materials of this earth occupied a volume thousands of times greater than they do at present.

In a previous chapter we have seen how the sun was at one time in the nebulous state, and now we have been led to a similar conclusion with regard to the earth. At that time, of course, the sun was greatly in excess of its present dimensions, and the earth was also greatly swollen. The nebula which formed our sun, and the nebula which formed our earth, were both so vast as to be confluent; they were indeed both part of the same vast nebula.

Such has been the Earth's Beginning so far as modern science can make it clear to us. We have at least indicated the course which events must have taken according to the laws of nature as we understand them. Many of the details of the great evolution are no doubt unknown at present, and perhaps must ever remain so. That the events which we have endeavoured to describe do substantially represent the actual evolution of our system is the famous Nebular Theory.

Chapter 9

EARTHQUAKES AND VOLCANOES

In this chapter we shall learn what we can as to the physical condition of the interior of our earth so far as it may be reasonably inferred from the facts of observation. We have already explained in the last chapter that a very high temperature must be found at the depth of even a small fraction of the earth's radius, and we have pointed out that the excessively high pressure characteristic of the earth's interior must be borne in mind in any consideration as to the condition of the matter there found.

Let us take, for instance, that primary question in terrestrial physics, as to whether the interior of the earth is liquid or solid. If we were to judge merely from the temperatures reasonably believed to exist at a depth of some twenty miles, and if we might overlook the question of pressure, we should certainly say that the earth's interior must be in a fluid state. It seems at least certain that the temperatures to be found at depths of two score miles, and still more at greater depths, must be so high that the most refractory solids, whether metals or minerals, would at once yield if we could subject them to such temperatures in our laboratories. At such temperatures every metal would become fluid, even if it were not transformed into a cloud of vapour. But none of our laboratory experiments can tell us whether, under the pressure of thousands of tons on the square inch, the application of any heat whatever would be adequate to transform solids into liquids. It may indeed be reasonably doubted whether the terms solids and liquids are applicable, in the sense in which we understand them, to the materials forming the interior of the earth.

It was my good fortune some years ago to enjoy a most interesting trip to Norway, in company with a distinguished geologist. Under his guidance I there saw evidence which demonstrates conclusively that, when subjected to great pressure, solids, as we should call them, behave in a manner which, if not that of actual liquids, resembles at all events in some of its characteristics the behaviour of liquids. These rocks in some places are conglomerates, of which the leading constituents are water-worn pebbles of granite. These pebbles are of various sizes, from marbles to paving-stones. In some parts of the country these granite pebbles remain in the form which they acquired on the beach on which they were rolled by the primæval ocean; in other parts of the same interesting region the form of the pebbles has been greatly changed from what it was originally. For in the course of geological periods, and after the pebbles had become consolidated into the conglomerate, the rock so formed had been in some cases submitted to enormous pressure. This may have been lateral pressure, such as is found to have occurred in many other places, where it has produced the well-known geological phenomenon of strata crumpled into folds. In the present case, however, it seemed more probable that it was the actual weight of the superincumbent rocks, which once lay over these beds of conglomerate, which produced the surprising transformation. It seems to be not at all improbable that at one time these beds of conglomerate must have been covered with strata of which the thickness is so great that it may actually be estimated by miles. There has, however, been immense denudation of the superficial rocks in this part, at all events, of Norway, so that in the course of ages these strata, overlying the conglomerate for ages, have been so far worn away, and indeed removed, by the action of ice and the action of water that the conglomerate is now exposed to view. It offers for our examination striking indications of the enormous pressure to which it was subjected during the incalculable ages of geological time.

The effect of this long continuance of great pressure upon the pebbles of the conglomerate in certain parts of the country has been most astonishing. The granite in the pebbles still retains its characteristic crystalline structure; it has obviously not undergone anything that could be described as fusion; yet under the influence of the two factors of that pressure, namely, its intensity and its long continuance, the granite pebbles have yielded. In some cases they are slightly elongated, in others they are much elongated, while in yet others they are even rolled out flat. At different places along the valley the various phases of the transformation can be studied. We can find places where the pebbles seem little altered, and then we can trace each stage until the solid granite pebbles have, by the application of excessive pressure, been compressed into thin sheets whose character it would not have been easy to divine if it had not been possible to trace out their history. These sheets lie close and parallel, so that the material thus produced acquires some of the characteristics of slate. It splits easily along the flattened sheets, and this rolled-out conglomerate is indeed actually used as a substitute for slate, and in some places there are houses roofed with the conglomerate which has been treated in this extraordinary fashion.

This fact will illustrate a principle, already well known in the arts, that many, if not all, solids may be made to flow like liquids if only adequate pressure be applied. The making of lead tubes is a well-known practical illustration of the same principle, for these tubes are simply formed by forcing solid lead by the hydraulic press through a mould which imparts the desired form.

If then a solid can be made to behave like a liquid, even with such pressures as are within our control, how are we to suppose that the solids would behave with such pressures as those to which they are subjected in the interior of the earth? The fact is that the terms solid and liquid, at least as we understand them, appear to have no physical meaning with regard to bodies subjected to these stupendous pressures, and this must be carefully borne in mind when we are discussing the nature of the interior of the earth.

It must, however, be admitted that the interior of the earth in its actual physical state seems to possess at least one of the most important characteristics of a solid, for it seems to be intensely rigid. We mean by this, that the material of the earth, or rather each particle of that material, is very little inclined to move from its position with reference to the adjacent particles by the application of force. Possibly a liquid, such as water, might not behave very differently in this respect from a solid such as cast iron, if each of them were exposed to a pressure of scores of thousands of tons per square inch, as are the materials which form the great bulk of the earth. But, without speculating on these points, we are able to demonstrate that the earth, as a whole, does exhibit extreme rigidity. This is one of the most remarkable discoveries which has ever been made with regard to the physics of our earth. The discovery that the earth is so rigid is mainly due to Lord Kelvin.

We shall now mention the line of evidence which appears to prove, in the simplest and most direct manner, the excessive rigidity of our earth. It is derived from the study of earthquake phenomena, and we must endeavour to set it forth with the completeness its importance deserves.

As to the immediate cause of earthquakes, there is no doubt considerable difference of opinion. But I think it will not be doubted that an earthquake is one of the consequences, though perhaps a remote one, of the gradual loss of internal heat from the earth. As this terrestrial heat is gradually declining, it follows from the law that we have already so often had occasion to use that the bulk of the earth must be shrinking. No doubt the diminution in the earth's diameter, due to the loss of heat must be excessively small, even in a long period of time. The cause, however, is continually in operation, and accordingly the crust of the earth has, from time to time, to be accommodated to the fact that the whole globe is lessening. The circumference of our earth at the Equator must be gradually declining; a certain length in that circumference is lost each year. We may admit that loss to be a quantity far too small to be measured by any observations as yet obtainable, but, nevertheless, it is productive of phenomena so important that it cannot be overlooked.

It follows from these considerations that the rocks which form the earth's crust over the surface of the continents and the islands, or beneath the beds of ocean, must have a lessening acreage year by year. These rocks must therefore submit to compression, either continuously or from time to time, and the necessary yielding of the rocks will in general take place in those regions where the materials of the earth's crust happen to have comparatively small powers of resistance. The acts of compression will often, and perhaps generally, not proceed with uniformity, but rather with small successive shifts, and even though the displacements of the rocks in these shifts be actually very small, yet the pressures to which the rocks are subjected are so vast that a very small shift may correspond to a very great terrestrial disturbance.

Suppose, for instance, that there is a slight shift in the rocks on each side of a crack, or fault, at a depth of ten miles. It must be remembered that the pressure ten miles down would be about thirty-five tons on the square inch. Even a slight displacement of one extensive surface over another, the sides being pressed together with a force of thirty-five tons on the square inch, would be an operation necessarily accompanied by violence greatly exceeding that which we might expect from so small a displacement if the forces concerned had been only of more ordinary magnitude. On account of this great multiplication of the intensity of the phenomenon, merely a small rearrangement of the rocks in the crust of the earth, in pursuance of the necessary work of accommodating its volume to the perpetual shrinkage, might produce an excessively violent shock extending far and wide. The effect of such a shock would be propagated in the form of waves through the globe, just as a violent blow given at one end of a bar of iron by a hammer is propagated through the bar in the form of waves. When the effect of this internal adjustment reaches the earth's surface, it will sometimes be great enough to be perceptible in the shaking it gives that surface. The shaking may be so violent that buildings may not be able to withstand it. Such is the phenomenon of an earthquake.

Earthquakes have been made to yield testimony of the most striking character with regard to the rigidity of the earth. The researches we are now to describe are mainly due to Professor Milne, who, having enjoyed the advantage of studying earthquakes in their natural home in Japan, where are to be found some of the most earthquake-shaken regions of this earth, has now transferred his observations of these phenomena to the more peaceful regions of the Isle of Wight. But though the Isle of Wight is perhaps one of the last places in the world to which anyone who desired to experience violent earthquake shocks would be likely to go, yet by the help of a beautiful apparatus Professor Milne is actually able to witness important earthquakes that are happening all over the world. He has a demonstration of these earthquakes in the indications of an extremely sensitive instrument which he has erected in his home at Shide.

When our earth is shaken by one of those occasional adjustments of the crust which I have described, the wave that spreads like a pulsation from the centre of agitation extends all over our globe and, indeed I may say, is transmitted right through it. At the surface lying immediately over the centre of disturbance there will be a violent shock. In the surrounding country, and often over great distances, the earthquake may also be powerful enough to produce destructive effects. The convulsion may also be manifested over a far larger area of country in a way which makes the shock to be felt, though the damage wrought may not be appreciable. But beyond a limited distance from the centre of the agitation the earthquake will produce no destructive effects upon buildings, and will not even cause vibrations that would be appreciable to ordinary observation.

This earth of ours may transmit from an earthquake pulses of a very distinct and definite character, which are too weak to be perceived by our unaided senses; but, just as the microscope will render objects visible which are too minute to be perceived without this aid to the ordinary vision, so these faint earth-pulses may be rendered perceptible by the delicate indications of an instrument which perceives and records tremors that would pass unnoticed by our ordinary observations. The ingenious instrument for studying earthquakes is called a seismometer. It marks on a revolving drum of paper the particulars of those infinitesimal tremors by which the earth is almost daily agitated in one place or another.

Let us suppose, for example, that an earthquake occurs in Japan, in which much agitated country it is, I believe, estimated that no fewer than one thousand earthquakes of varying degrees of intensity occur annually in one district or another. Let us suppose that this earthquake behaves as serious earthquakes usually do; that it knocks down buildings and monuments, causes landslips, raises great waves in the sea and hurls them as inundations on the land. We may also suppose that it issues tragically in the loss of many lives and that there is a destruction of much property, and that its energies in the acutely violent form extend over, let us say, an area of a hundred square miles. Beyond that area of greatest destruction such an earthquake would be felt over a great extent of country as a shaking more or less vehement, and characteristic rumbling sounds would be heard. But the intensity declines with the distance, and we may feel confident that not even the faintest indications of the earthquake would be perceptible by the unaided senses at a thousand miles from its origin. A thousand miles is, however, less than a fifth of the distance between Tokio and Shide, in the Isle of Wight, measured in a great circle round the earth's surface. The acutest sense could not perceive the slightest indication of the convulsion in Japan at even half the distance between these two places. But the earth transmits so faithfully the undulations committed to its care that though the intensity may have declined so as to be no longer perceptible to the unaided sense, it is still possible that they may be shown distinctly on the seismometer in Professor Milne's laboratory, even after a journey of five thousand miles. This instrument not only announces that an earthquake has been in progress some little time previously, but the recording pencil reproduces with marvellous fidelity some actual details of the vibration. The movements of the line up and down on the revolving drum of paper show how the convulsions succeed each other, and their varying intensity. Thus Professor Milne is enabled to set down some features of the earthquake long before the post brings an account of the convulsion from the unhappy locality.

Professor Milne's account of work in studying earthquakes has the charm of a romance, even while it faithfully sets out the facts of Nature. I have supposed the earthquake to take place in Japan; but we must observe that the seismometer at Shide will also take account of considerable earthquakes in whatever part of the world the disturbance may arise. There are, for example, localities in the West Indies in which earthquakes are by no means infrequent, though they may not be phenomena of almost daily occurrence, as they are in Japan. Every considerable earthquake, no matter where its centre may lie, produces in our whole globe a vibration or a tingle which is sufficient to be manifested by the delicate indications of the seismometer at Shide. Thus this instrument, which in the morning may record an earthquake from Japan, will in the afternoon of the same day delineate with equal fidelity an earthquake from the opposite hemisphere in the neighbourhood of the Caribbean Sea.

In each locality in which earthquakes are chronic it would seem as if there must be some particularly weak spot in the earth some miles below the surface. A shrinkage of the earth, in the course of the incessant adjustment between the interior and the exterior, will take place by occasional little jumps at this particular centre. The fact that there is this weak spot at which small adjustments are possible may provide, as it were, a safety-valve for other places in the same part of the world. Instead of a general shrinking, the materials would be sufficiently elastic and flexible to allow the shrinking for a very large area to be done at this particular locality. In this way we may explain the fact that immense tracts on the earth are practically free from earthquakes of a serious character, while in the less fortunate regions the earthquakes are more or less perennial.

The characteristics of an earthquake record, a *seismogram*, if we give it the correct designation, depend on the distance of the origin from the locality where the record is made. The length of the journey, as might be expected, tells on the character of the inscription which the earthquake waves make on the drum. If, for instance, the first intimation of a large earthquake received at Shide precedes the second by about thirty-five minutes, it may be concluded that the earthquake has come from Japan.

In like manner the shocks, with their origin in the West Indies, will proceed from their particular earthquake centre, and consequently all the earthquakes from this source will possess a characteristic resemblance. The Japan group of earthquakes will have, so to speak, a family resemblance; and the Trinidad group of earthquakes, though quite different from the Japan group, will also possess a family resemblance. These features are faithfully transmitted by undulations through the earth and round the earth; thus in due course they reach the Isle of Wight, and they are reproduced by the pencil of the seismometer. The different earthquakes of a family may differ in size, in intensity, and undulation, but they will have the features appropriate to the particular group from which they come. From long experience Professor Milne has become so familiar with the lineaments of these earthquake families, that in his study at Shide, as he looks at the indications of his instrument, he is able to say, for example, "Here is an earthquake, and it is a little earthquake from Japan;" then a little later, when a new earthquake begins, he will say, "And here is a big earthquake from Trinidad."

Professor Milne's apparatus has brought us remarkable information with regard to the interior of the earth. The story which we have to tell is really one of the most astonishing in physical science. Let us suppose that an earthquake originates in Japan. We shall assume that the earthquake is a vigorous one, capable of producing bold and definite indications on the seismometer even in the Isle of Wight. It is to be noted that this instrument is not content merely with a single version of the story of that earthquake; it will indeed repeat that story twice more. First of all, about a quarter of an hour after a shock has taken place in Japan, the pencil of the seismometer commences to record. But this record, though quite distinct, is not so boldly indicated as the subsequent records of the same event which will presently be received. It is to be regarded as a precursor. After the first record is completed there is a pause of perhaps three-quarters of an hour, and then the pencil of the seismometer commences again. It commences to give an earthquake record, but it is obviously only a second version of the same earthquake. For the ups and downs traced by the pencil are just the same relatively as before. The picture given of the earthquake is, however, on a much larger scale than the one that is first sent. The extent of the shaking of the instrument in this second record is greater than in the first, and all the details are more boldly drawn.

After the second diagram has been received, there is yet another pause, which may be perhaps for half an hour. Then, by the same pencil, a third and last version is conveyed to the seismometer. This diagram is not quite so strong as the last, though stronger than the first; in it again, however, the faithful pencil tells, with many a detail, what happened in this earthquake at Japan.

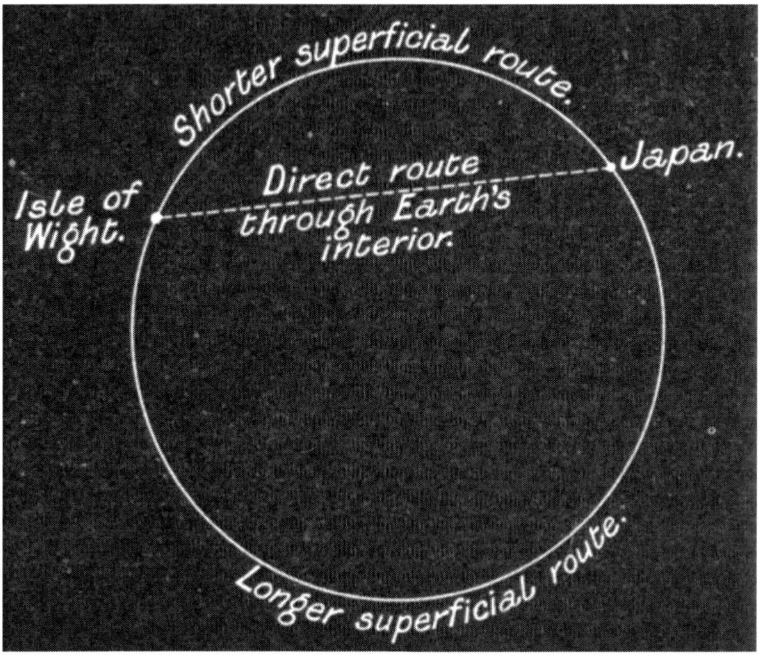

Figure 26. Earthquake Routes from Japan to the Isle of Wight.

We have first to explain how it occurs that there are three versions of the event, for it need hardly be said that the same earthquake did not take place three different times over. The point is indeed a beautiful one. The explanation is so astonishing that we should hardly credit it were it not established upon evidence that does not admit of a moment's question.

In the adjoining diagram we represent the position of Japan at one side of the earth, and the Isle of Wight at the other. When the earthquake takes place at Japan it originates, as we have said, a series of vibrations through our globe. We must here distinguish between the rocks—I might almost say the comparatively pliant rocks—which form the earth's crust, and those which form the intensely rigid core of the interior of our globe. The vibrations which carry the tidings of the earthquake spread through the rocks on the surface, from the centre of the disturbance, in gradually enlarging circles. We may liken the spread of these vibrations to the ripples in a pool of water which diverge from the spot where a raindrop has fallen, or to the remarkable air-waves from Krakatoa, to which we shall presently refer. The vibrations transmitted by the rocks on the surface, or on the floor of the ocean, will carry the message all over the earth. As these rocks are flexible, at all events by comparison with the earth's interior, the vibrations will be correspondingly large, and will travel with vigour over land and under sea. In due time they reach the Isle of Wight, where they set the pencil of the seismometer at work. But there are different ways round the earth from Japan to the Isle of Wight. There is the most direct route across Asia and Europe; there is also the route across the Pacific, America, and the Atlantic. The vibrations will travel by both routes, and the former is the shorter of the two. The vibrations which take the first route through the crust of the earth's surface are travelling by the shorter distance; they consequently reach Shide first, and render their version of what has happened. But the vibrations which, starting from the centre of the disturbance, move through the earth's crust in an opposite direction will also in their due course of expansion reach the Isle of Wight. They will have had a longer journey, and will consequently be somewhat enfeebled, though they will still retain the characteristics marking the particular earthquake centre from which they arose.

We thus account for both the second and the third of the different versions of the earthquake which are received at Shide. And now for the first of the three versions. This is the one which is of special interest to us at present. The original subterranean impulse was, as we have seen, propagated through the rocks forming the earth's crust. Part of it, however, entered into the core forming the earth's interior. The earthquake had the power not only of shaking the earth's crust all over, but it produced the astonishing effect of setting the whole interior of our globe into a tremble. There was not a single particle of our earth, from centre to surface, which was not made to vibrate, in some degree, in consequence of the earthquake. Certain of these vibrations, spreading from the centre of disturbance, took a direct course to the Isle of Wight, right through the globe. They consequently had a shorter journey in travelling from Tokio to Shide than those which went round the earth's crust. The former travelled near the chord, while the latter travelled on the arc.

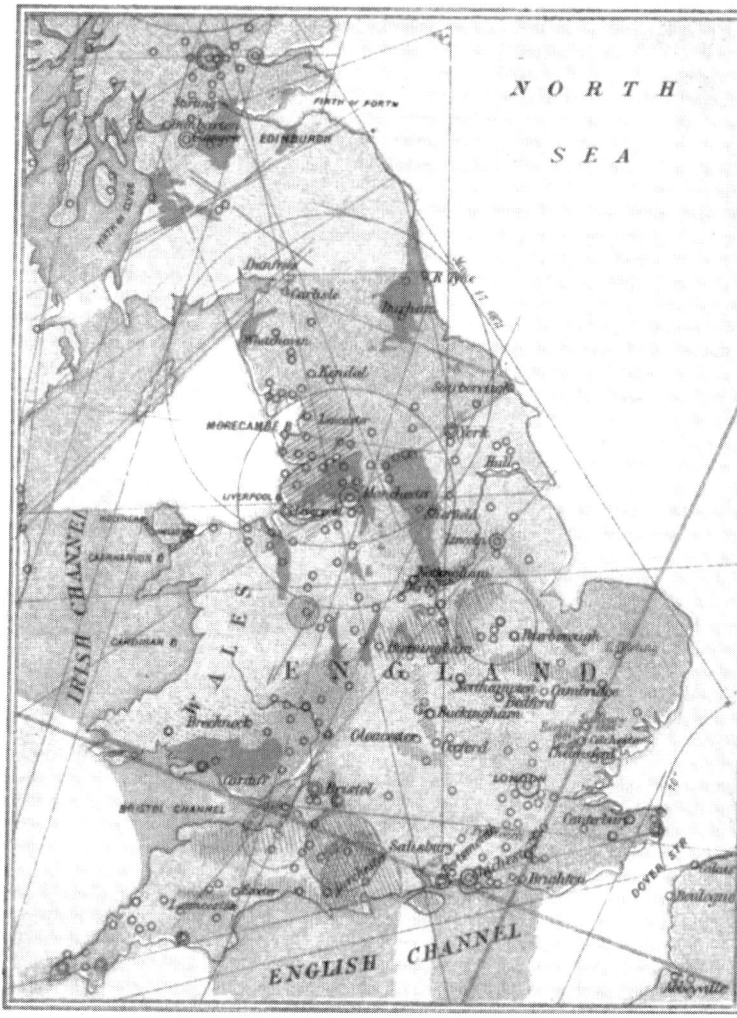

Figure 27. Showing Localities of Earthquakes.

Even for this reason alone the internal vibrations might be expected to accomplish their journey more rapidly than the superficial movements. With the same velocity they would take a shorter time for the journey. There is, however, another reason for the lesser time taken by the internal vibrations. Not only is the journey shorter, but the speed with which these vibrations travel through the solid earth is much greater than the speed with which superficial vibrations travel through the crust. It has been shown that the average velocity of these vibrations when travelling through the centre of the earth is rather more than ten miles a second. The velocity varies with the square root of the depth, and near the surface it is scarcely two miles a second. There are two points to be specially noticed. The vibrations, which, passing through the earth's interior with a high velocity, arrive as precursors, make a faithful diagram, but only on a very small scale. We say that these vibrations have but small amplitude. This shows that the particles in the earth's interior are not much displaced by the earthquake, as compared with those on the earth's crust, and this is one indication of the effective rigidity of the earth. It is also to be noted that the great speed with which the vibrations traverse the solid earth is a consequence of the extreme rigidity of our globe. These vibrations travel more rapidly through the earth than they would do through a bar of solid steel. In other words, we have here a proof that, under the influence of the tremendous pressures characteristic of the earth's interior, the material of which that earth is composed, notwithstanding the high temperature to which it is raised, possesses a rigidity which is practically greater than that of steel itself.

This is perhaps the most striking testimony that can be borne to the rigidity of our globe; but we must not imagine that we are dependent solely upon the phenomena of earthquakes for the demonstration of this important point; there are other proofs. It can be shown that the ebb and flow of the tides on our coasts would be very different from that which they actually are were it not that the earth behaves as a rigid globe. It has also been demonstrated that certain astronomical phenomena connected with the way in which the earth turns round on its axis would not be the same as we actually find them to be if the earth were not solid in its interior.

The result of these investigations is to show that, though this globe of ours must be excessively hot inside, so hot indeed that at ordinary pressures even the most refractory solids would be liquefied or vaporised, yet under the influence of the pressure to which its materials are subjected the behaviour of that globe is as that of the most rigidly solid body.

Happily in this country we do not often experience earthquakes other than delicate movements shown by the record of the seismometer. But though most of us live our lives without ever having felt an earthquake shock, yet earthquakes do sometimes make themselves felt in Great Britain. The map we here give, which was drawn by Professor J. P. O'Reilly, indicates the localities in England in which from time to time earthquake shocks have been experienced.

The internal heat of the earth, derived from the primæval nebula, is in no way more strikingly illustrated than by the phenomena of volcanoes. We have shown in this chapter that there is no longer any reason to believe that the earth is fluid in its interior. The evidence has proved that, under the extraordinary pressure which prevails in the earth, the materials in the central portions of our globe behave with the characteristics of solids rather than of liquids. But though this applies to the deep-seated regions of our globe, it need not universally apply at the surface or within a moderate depth from the surface. When the circumstances are such that the pressure is relaxed, then the heat is permitted to exercise its property of transforming the solids into liquids. Masses of matter near the earth's crust are thus, in certain circumstances, and in certain localities, transformed into the fluid or viscid form. In that state they may issue from a volcano and flow in sluggish currents as lava.

There has been much difference of opinion as to the immediate cause of volcanic action, but there can be little doubt that the energy which is manifested in a volcanic eruption has been originally derived in some way from the contraction of the primæval nebula. The extraordinary vehemence that a volcanic eruption sometimes attains may be specially illustrated by the case of the great eruption of Krakatoa. It is, indeed, believed that in the annals of our earth there has been no record of a volcanic eruption so vast as that which bears the name of this little island in far Eastern seas, ten thousand miles from our shores.

Until the year 1883 few had ever heard of Krakatoa. It was unknown to fame, as are hundreds of other gems of glorious vegetation set in tropical waters. It was not inhabited, but the natives from the surrounding shores of Sumatra and Java used occasionally to draw their canoes up on its beach, while they roamed through the jungle in search of the wild fruits that there abounded. Geographers in early days hardly condescended to notice Krakatoa; the name of the island on their maps would have been far longer than the island itself. It was known to the mariner who navigated the Straits of Sunda, for it was marked on his charts as one of the perils of the intricate navigation in those waters. It was no doubt recorded that the locality had been once, or more than once, the seat of an active volcano. In fact, the island seemed to owe its existence to some frightful eruption of bygone days; but for a couple of centuries there had been no fresh outbreak. It almost seemed as if Krakatoa might be regarded as a volcano that had become extinct. In this respect it would only be like many other similar objects all over the globe, or the countless extinct volcanoes all over the moon.

In 1883 Krakatoa suddenly sprang into notoriety. Insignificant though it had hitherto seemed, the little island was soon to compel by its tones of thunder the whole world to pay it instant attention. It was to become the scene of a volcanic outbreak so appalling that it is destined to be remembered throughout the ages. In the spring of that year there were symptoms that the volcanic powers in Krakatoa were once more about to awake from the slumber that had endured for many generations. Notable warnings were given. Earthquakes were felt, and deep rumblings proceeded from the earth, showing that some disturbance was in preparation, and that the old volcano was again to burst forth after its long period of rest. At first the eruption did not threaten to be of any serious type; in fact, the good people of Batavia, so far from being terrified at what was in progress in Krakatoa, thought the display was such an attraction that they chartered a steamer and went forth for a pleasant picnic to the island. Many of us, I am sure, would have been delighted to have been able to join the party who were to witness so interesting a spectacle. With cautious steps the more venturesome of the excursion party clambered up the sides of the volcano, guided by the sounds which were issuing from its summit. There they beheld a vast column of steam pouring forth with terrific noise from a profound opening about thirty yards in width.

As the summer of this dread year advanced the vigour of Krakatoa steadily increased, the noises became more and more vehement; these were presently audible on shores ten miles distant, and then twenty miles distant; and still those noises waxed louder and louder, until the great thunders of the volcano, now so rapidly developing, astonished the inhabitants that dwelt over an area at least as large as Great Britain. And there were other symptoms of the approaching catastrophe. With each successive convulsion a quantity of fine dust was projected aloft into the clouds. The wind could not carry this dust away as rapidly as it was hurled upwards by Krakatoa, and accordingly the atmosphere became heavily charged with suspended particles. A pall of darkness thus hung over the adjoining seas and islands. Such was the thickness and the density of these atmospheric volumes of Krakatoa dust that, for a hundred miles around, the darkness of midnight prevailed at midday. Then the awful tragedy of Krakatoa took place. Many thousands of the unfortunate inhabitants of the adjacent shores of Sumatra and Java were destined never to behold the sun again. They were presently swept away to destruction in an invasion of the shore by the tremendous waves with which the seas surrounding Krakatoa were agitated.

Gradually the development of the volcanic energy proceeded, and gradually the terror of the inhabitants of the surrounding coasts rose to a climax. July had ended before the manifestations of Krakatoa had attained their full violence. As the days of August passed by the spasms of Krakatoa waxed more and more vehement. By the middle of that month the panic was widespread, for the supreme catastrophe was at hand.

On the night of Sunday, August 26th, 1883, the blackness of the dust-clouds, now much thicker than ever in the Straits of Sunda and adjacent parts of Sumatra and Java, was only occasionally illumined by lurid flashes from the volcano. The Krakatoan thunders were on the point of attaining their complete development. At the town of Batavia, a hundred miles distant, there was no quiet that night. The houses trembled with the subterranean violence, and the windows rattled as if heavy artillery were being discharged in the streets. And still these efforts seemed to be only rehearsing for the supreme display. By ten o'clock on the morning of Monday, August 27th, 1883, the rehearsals were over and the performance began. An overture, consisting of two or three introductory explosions, was succeeded by a frightful convulsion which tore away a large part of the island of Krakatoa and scattered it to the winds of heaven. In that final effort all records of previous explosions on this earth were completely broken.

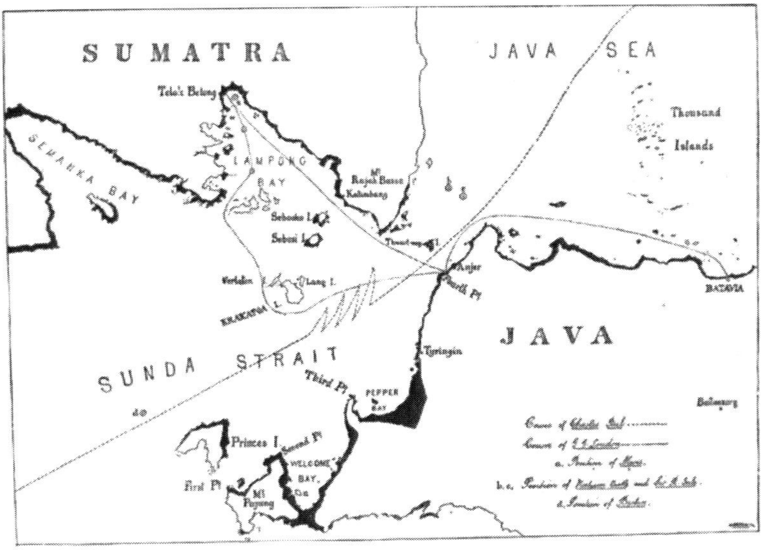

Figure 28. Showing Coasts invaded by the Great Sea-waves from Krakatoa. (*From the Royal Society's Reports.*)

This supreme effort it was which produced the mightiest noise that, so far as we can ascertain, has ever been heard on this globe. It must have been indeed a loud noise which could travel from Krakatoa to Batavia and preserve its vehemence over so great a distance; but we should form a very inadequate conception of the energy of the eruption of Krakatoa if we thought that its sounds were heard by those merely a hundred miles off. This would be little indeed compared with what is recorded, on testimony which it is impossible to doubt.

Westward from Krakatoa stretches the wide expanse of the Indian Ocean. On the opposite side from the Straits of Sunda lies the island of Rodriguez, the distance from Krakatoa being almost three thousand miles. It has been proved by evidence which cannot be doubted that the booming of the great volcano attracted the attention of an intelligent coastguard on Rodriguez, who carefully noted the character of the sounds and the time of their occurrence. He had heard them just four hours after the actual explosion, for this is the time the sound occupied on its journey.

We shall better realise the extraordinary vehemence of this tremendous noise if we imagine a similar event to take place in localities more known to most of us than are the far Eastern seas.

Figure 29. The early stage of the Eruption of Krakatoa.
(*From a Photograph taken on May 27th, 1883.*)

If Vesuvius were vigorous enough to thunder forth like Krakatoa, how great would be the consternation of the world! Such a report might be heard by King Edward at Windsor, and by the Czar of all the Russias at Moscow. It would astonish the German Emperor and all his subjects. It would penetrate to the seclusion of the Sultan at Constantinople. Nansen would still have been within its reach when he was furthest north, near the Pole. It would have extended to the sources of the Nile, near the Equator. It would have been heard by Mohammedan pilgrims at Mecca. It would have reached the ears of exiles in Siberia. No inhabitant of Persia would have been beyond its range, while passengers on half the liners crossing the Atlantic would also catch the mighty reverberation.

The subject is of such exceptional interest that I may venture on another illustration. Let us suppose that a similar earth-shaking event took place in a central position in the United States. Let us say, for example, that an explosion occurred at Pike's Peak as resonant as that from Krakatoa. It would certainly startle not a little the inhabitants of Colorado far and wide. The ears of dwellers in the neighbouring States would receive a considerable shock. With lessening intensity the sound would spread much further around—indeed, it might be heard all over the United States. The sonorous waves would roll over to the Atlantic coast, they would be heard on the shores of the Pacific. Florida would not be too far to the south, nor Alaska too remote to the north. If, indeed, we could believe that the sound would travel as freely over the great continent as it did across the Indian Ocean, then we may boldly assert that every ear in North America might listen to the thunder from Pike's Peak, if it rivalled Krakatoa. The reverberation might even be audible by skin-clad Eskimos amid the snows of Greenland, and by naked Indians sweltering on the Orinoco. Can we doubt that Krakatoa made the greatest noise that has ever been recorded?

Among the many other incidents connected with this explosion, I may specially mention the wonderful system of divergent ripples that started in our atmosphere from the point at which the eruption took place. I have called them ripples, from the obvious resemblance which they bear to the circular expanding ripples produced by raindrops which fall upon the still surface of water.

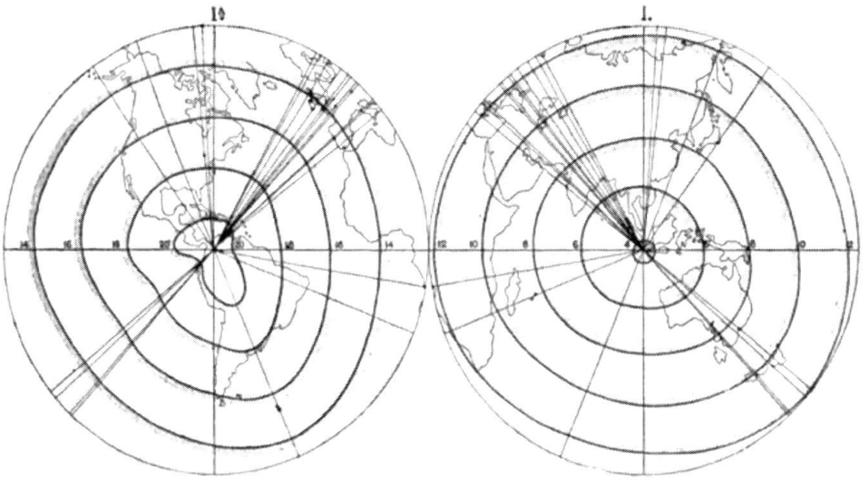

Figure 30. Spread of the Air-wave from Krakatoa to the Antipodes.
(*From the Royal Society's Reports.*)

But it would be more correct to say that these objects were a series of great undulations which started from Krakatoa and spread forth in ever-enlarging circles through our atmosphere. The initial impetus was so tremendous that these waves spread for hundreds and thousands of miles. They diverged, in fact, until they put a mighty girdle round the earth, on a great circle of which Krakatoa was the pole. The atmospheric waves, with the whole earth now well in their grasp, advanced into the opposite hemisphere. In their further progress they had necessarily to form gradually contracting circles, until at last they converged to a point in Central America, at the very opposite point of the diameter of our earth, eight thousand miles from Krakatoa. Thus the waves completely embraced the earth. Every part of our atmosphere had been set into a tingle by the great eruption. In Great Britain the waves passed over our heads, the air in our streets, the air in our houses, trembled from the volcanic impulse. The very oxygen supplying our lungs was responding also to the supreme convulsion which took place ten thousand miles away. It is needless to object that this could not have taken place because we did not feel it. Self-registering barometers have enabled these waves to be followed unmistakably all over the globe.

Such was the energy with which these vibrations were initiated at Krakatoa, that even when the waves thus arising had converged to the point diametrically opposite in South America their vigour was not yet exhausted. The waves were then, strange to say, reflected back from their point of convergence to retrace their steps to Krakatoa. Starting from Central America, they again described a series of enlarging circles, until they embraced the whole earth. Then, advancing into the opposite hemisphere, they gradually contracted until they had regained the Straits of Sunda, from which they had set forth about thirty-six hours previously. Here was, indeed, a unique experience. The air-waves had twice gone from end to end of this globe of ours. Even then the atmosphere did not subside until, after some more oscillations of gradually fading intensity, at last they became evanescent.

But, besides these phenomenal undulations, this mighty incident at Krakatoa has taught us other lessons on the constitution of our atmosphere. We previously knew little, or I might almost say nothing, as to the conditions prevailing above the height of ten miles overhead. We were almost altogether ignorant of what the wind might be at an altitude of, let us say, twenty miles. It was Krakatoa which first gave us a little information which was greatly wanted. How could we learn what winds were blowing at a height four times as great as the loftiest mountain on the earth, and twice as great as the loftiest altitude to which a balloon has ever soared? We could neither see these winds nor feel them. How, then, could we learn whether they really existed? No doubt a straw will show the way the wind blows, but there are no straws up there. There was nothing to render the winds perceptible until Krakatoa came to our aid. Krakatoa drove into those winds prodigious quantities of dust. Hundreds of cubic miles of air were thus deprived of that invisibility which they had hitherto maintained. They were thus compelled to disclose those movements about which, neither before nor since, have we had any opportunity of learning.

With eyes full of astonishment men watched those vast volumes of Krakatoa dust start on a tremendous journey. Westward the dust of Krakatoa took its way. Of course, everyone knows the so-called tradewinds on our earth's surface, which blow steadily in fixed directions, and which are of such service to the mariner. But there is yet another constant wind. We cannot call it a trade-wind, for it never has rendered, and never will render, any service to navigation. It was first disclosed by Krakatoa. Before the occurrence of that eruption no one had the slightest suspicion that far up aloft, twenty miles over our heads, a mighty tempest is incessantly hurrying with a speed much greater than that of the awful hurricane which once laid so large a part of Calcutta on the ground, and slew so many of its inhabitants. Fortunately for humanity, this new trade-wind does not come within less than twenty miles of the earth's surface. We are thus preserved from the fearful destruction that its unintermittent blasts would produce, blasts against which no tree could stand, and which would, in ten minutes, do as much damage to a city as would the most violent earthquake. When this great wind had become charged with the dust of Krakatoa, then, for the first and, I may add, for the only time, it stood revealed to human vision. Then it was seen that this wind circled round the earth in the vicinity of the Equator, and completed its circuit in about thirteen days.

Please observe the contrast between this wind of which we are now speaking and the waves to which we have just referred. The waves were merely undulations or vibrations produced by the blow which our atmosphere received from the explosion of Krakatoa, and these waves were propagated through the atmosphere much in the same way as sound waves are propagated. Indeed, these waves moved with the same velocity as sound. But the current of air of which we are now speaking was not produced by Krakatoa; it existed from all time, before Krakatoa was ever heard of, and it exists at the present moment, and will doubtless exist as long as the earth's meteorological arrangements remain as they are at present. All that Krakatoa did was simply to provide the charges of dust by which for one brief period this wind was made visible.

In the autumn of 1883 the newspapers were full of accounts of strange appearances in the heavens. The letters containing these accounts poured in upon us from residents in Ceylon; they came from residents in the West Indies, and from other tropical places. All had the same tale to tell. Sometimes experienced observers assured us that the sun looked blue; sometimes we were told of the amazement with which people beheld the moon draped in vivid green. Other accounts told of curious halos, and, in short, of the signs in the sun, the moon, and the stars, which were exceedingly unusual, even if we do not say that they were absolutely unprecedented.

Those who wrote to tell of the strange hues that the sun manifested to travellers in Ceylon, or to planters in Jamaica, never dreamt of attributing the phenomena to Krakatoa, many thousands of miles away. In fact, these observers knew nothing at the time of the Krakatoa eruption, and probably few of them, if any, had ever heard that such a place existed. It was only gradually that the belief grew that these, phenomena were due to Krakatoa. But when the accounts were carefully compared, and when the dates were studied at which the phenomena were witnessed in the various localities, it was demonstrated that these phenomena, notwithstanding their worldwide distribution, had certainly arisen from the eruption in this little island in the Straits of Sunda. It was most assuredly Krakatoa that painted the sun and the moon, and produced the other strange and weird phenomena in the tropics.

After a little time we learned what had actually happened. The dust manufactured by the supreme convulsion was whirled round the earth in the mighty atmospheric current into which the volcano discharged it. As the dust-cloud was swept along by this incomparable hurricane, it showed its presence in the most glorious manner by decking the sun and the moon in hues of unaccustomed splendour and beauty. The blue colour in the sky under ordinary circumstances is due to particles in the air, and when the ordinary motes of the sunbeam were reinforced by the introduction of the myriads of motes produced by Krakatoa, even the sun itself sometimes showed a blue tint. Thus the progress of the great dust-cloud was traced out by the extraordinary sky effects it produced, and from the progress of the dust-cloud we inferred the movements of the invisible air current which carried it along. Nor need it be thought that the quantity of material projected from Krakatoa should have been inadequate to produce effects of this worldwide description. Imagine that the material which was blown to the winds of heaven by the supreme convulsion of Krakatoa could be all recovered and swept into one vast heap. Imagine that the heap were to have its bulk measured by a vessel consisting of a cube one mile long, one mile broad, and one mile deep; it has been estimated that even this prodigious vessel would have to be filled to the brim at least ten times before all the products of Krakatoa had been measured.

It was in the late autumn of 1883 that the marvellous series of celestial phenomena connected with the great eruption began to be displayed in Great Britain. Then it was that the glory of the ordinary sunsets was enhanced by a splendour which has dwelt in the memory of all those who were permitted to see them. The frontispiece of this volume contains a view of the sunset as seen at Chelsea at 4.40 p.m. on November 26th, 1883. The picture was painted from nature by Mr. W. Ascroft, and is given in the great work on Krakatoa which was published by the Royal Society. There is not the least doubt that it was the dust from Krakatoa which produced the beauty of those sunsets, and so long as that dust remained suspended in our atmosphere, so long were strange signs to be witnessed in the heavenly bodies. But the dust which had been borne with unparalleled violence from the interior of the volcano, the dust which had been shot aloft by the vehemence of the eruption to an altitude of twenty miles, the dust which had thus been whirled round and round our earth for perhaps a dozen times or more in this air current, which carried it round in less than a fortnight, was endowed with no power to resist forever the law of gravitation which bids it fall to the earth. It therefore gradually sank downwards. Owing, however, to the great height to which it had been driven, owing to the impetuous nature of the current by which it was hurried along, and owing to the exceedingly minute particles of which it was composed, the act of sinking was greatly protracted. Not until two years after the original explosion had all the particles with which the air was charged by the great eruption finally subsided on the earth.

At first there were some who refused to believe that the glory of the sunsets in London could possibly be due to a volcano in the Straits of

Sunda, at a distance from England which was but little short of that of Australia. But the gorgeous phenomena in England were found to be simultaneous with like phenomena in other places all round the earth. Once again the comparison of dates and other circumstances proved that Krakatoa was the cause of these exceptional and most interesting appearances.

Nor was the incident without a historical parallel, for has not Tennyson told us of the call to St. Telemachus—

> "Had the fierce ashes of some fiery peak
> Been hurl'd so high they ranged about the globe?
> For day by day, thro' many a blood-red eve,
> In that four-hundredth summer after Christ,
> The wrathful sunset glared...."

Chapter 10

SPIRAL AND PLANETARY NEBULÆ

In a great college in America a new educational experiment has been tried with some success. Instead of the instruction in history which students receive in most other institutions, an attempt has been made in this college to give instruction in a very different manner, which it is believed will not be of less educational value than the more ordinary processes of teaching. In the course of study to which I am now referring the student is invited to consider, not so much the history of the development of the Constitution of one particular country, as to make a broad survey of the different Constitutions under which the several countries of the world are at this moment governed. The promoters of this scheme believe that many of the intellectual advantages which are ordinarily expected to be gained by the study of the history of one country may be secured equally well by studying only existing conditions, provided that attention is given to several countries which have arrived at different stages of civilisation.

Without attempting to say how far the study of the existing Constitutions of France and Germany, America and Australia, Turkey and India, Morocco and Fiji, might be justly used to supersede the study of English history, it may at least be urged that if we had no annals from which history could be compiled it might be instructive to employ such a substitute for historical studies as is here suggested.

Figure 31. The Great Spiral Nebula (Lick Observatory).
(*From the Royal Astronomical Series.*)

This is, indeed, the course which we are compelled to take in our study of that great chapter in earth-history which we are discussing in these pages. It is obvious from the nature of the case that it can never be possible for us to obtain direct testimony as to what occurred in the bringing together of the materials of this globe. We must, therefore, look abroad through the universe, and see whether we can find, from the study of other systems at present in various stages of their evolution, illustrations of the incidents which we may presume to have occurred in the early stages of our own history. If Kant had never lived, if Laplace had never announced his Nebular Theory, if the discoveries of Sir William Herschel had not been made, I still venture to think that a due consideration of the remarkable photograph of the famous Great Spiral, which was obtained at the famous Lick Observatory in California, would have suggested the high probability of that doctrine which we describe as the Nebular Theory.

If an artist thoroughly versed in the great facts of astronomy had been commissioned to represent the nebular origin of our system as perfectly as a highly cultivated yet disciplined imagination would permit, I do not think he could have designed anything which could answer the purpose more perfectly than does that picture which is now before us. We might wish indeed that Kant and Laplace and Herschel could have lived to see this marvellous natural illustration of their views, for photographs were of course unthought of in those days, and, I need hardly say, that for any one celestial nebula that could have been known in the times of Laplace, hundreds are now within the reach of astronomers.

We entreat special attention to this picture which Nature has herself given us, and which represents what we may not unreasonably conclude to be a system in a state of formation. Let me say at once that our solar system, however imposing it may be from our point of view, is but of infinitesimal importance as compared with the system which is here in the course of development. It is sometimes urged that it is difficult to imagine how a system so large as ours could have been produced by condensation from a primæval nebula. The best answer is found in the fact that the Great Spiral now before us may be considered to exhibit at this very moment a system in actual evolution, the central body of which is certainly thousands of times, and not improbably millions of times, greater than the sun, and of which the attending planets or other revolving bodies, are framed on a scale immensely transcending that of even Jupiter himself. The details of this remarkable nebula seem to illustrate those particular features which had been previously assigned to the primæval nebula of our system, long before any photograph was available for the purpose of their study.

In the Great Nebula in Orion, to which we have already referred, as well as in many other similar objects which we might also have adduced, the nebulous material from which after long ages new systems may be the result, was shown in an extremely chaotic state. It was little more than an irregular stain of light on the sky. But in the picture of the Great Spiral which is before us it is manifest that the evolution of the system has reached an advanced stage; such considerable progress has been made in the actual formation that the final form seems to be shadowed forth. The luminosity is no longer diffused in a chaotic condition; it has formed into spirals, and become much condensed at the centre and somewhat condensed in other regions. As we now see it, the object seems to represent a system much more advanced in its formation than any of the other great nebulæ with which we have compared it. In comparison with it the evolution of such an object as the Great Nebula in Orion can hardly be said to have begun. But in the Great Spiral many portions of the nebula have already become outlined into masses which, though still far from resembling the planets in the solar system, have at least made some approach thereto while the central portions are being drawn together, just as we may conceive the great primæval fire-mist to have drawn together in the actual formation of the sun.

The famous nebula which we are discussing, and which is generally known as the Great Spiral, is found in the constellation of Canes Venatici, very near the end star in the tail of the Great Bear, and one-fourth of the way from it to Cor Caroli. It will be easy to find it from the indication given in the adjoining figure below. As a nebulous spot it is an object which can be seen with any moderately good telescope, but to detect those details which indicate the spiral structure demands an instrument of first-class power. This object had indeed been studied by many astronomers before Lord Rosse turned his colossal reflector upon it. Then it was that the wonderful whirlpool structure was first discovered, and thus the earliest spiral nebula became known.

Figure 32. How to Find the Great Spiral Nebula.

In those days there were few telescopes of great power, and none of those instruments appeared able to deal with this nebula sufficiently to reveal its spiral character. The announcement of the discovery of the spiral constitution of this object was therefore received with incredulity by some astronomers, who believed, or professed to believe, that the spiral lines of nebulous matter which Lord Rosse described so faithfully, existed only in the imagination of the astronomer. Indeed, in one notable instance, it was alleged that these features were to be attributed to actual imperfections in the unrivalled telescope. The incredulity widely prevalent in the middle of the last century about the existence of the spiral nebulæ may be paralleled by the incredulity about other discoveries in more recent years. When a highly skilled observer, using an instrument of adequate power, and, it may be, enjoying unequalled opportunities for good work, testifies to certain discoveries; when he has employed in the verification of his observations the skill and experience that years of practice have procured for him, it is futile for those who have not the like opportunities, either from the want of instruments of adequate power or from climatic difficulties, to deny the truth of discoveries because they are not able to verify them. It was absurd for astronomers to refuse assent to the great discoveries of Lord Rosse simply because instruments inferior to his would not show the spiral structure.

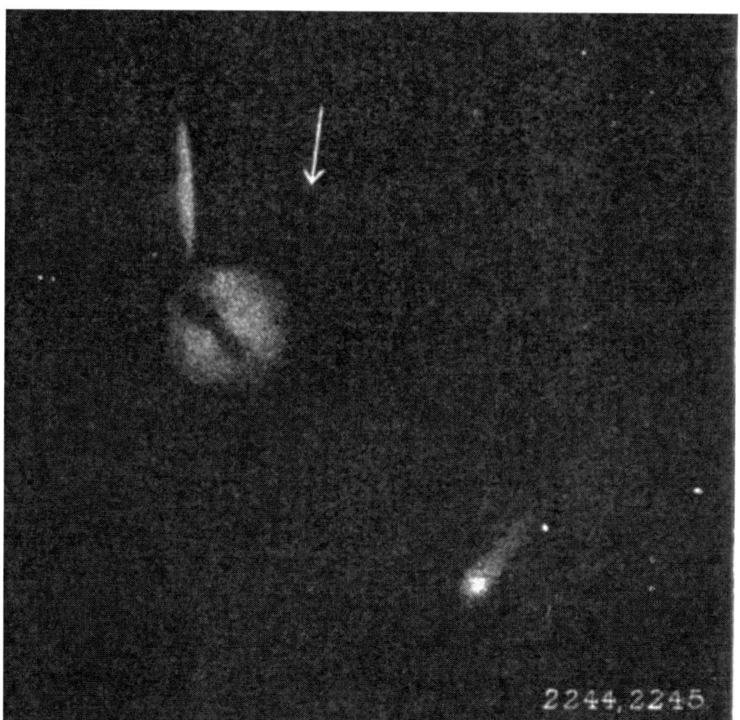

Figure 33. A Group of Nebulæ (Lord Rosse). (3440, 3445 in n.g.c.)
(From the Scientific Transactions of the Royal Dublin Society.)

In due time, one astronomer after another began to admit that possibly the remarkable form which Lord Rosse announced as characteristic of some nebulæ might not be a mere figment of the imagination. The complete vindication of Lord Rosse's great discovery was not, however, attained until that wonderful advance in the arts of astronomy when the photographic plate was called in to supplement, or rather vastly to extend, the powers of the eye. Dr. Isaac Roberts not only showed by a magnificent photograph that the Great Spiral discovered by Lord Rosse was just as Lord Rosse had described it, he not only showed that the other spirals announced by Lord Rosse were equally entitled to the name, but, with the newly acquired powers that the photographic plate placed at his disposal, he was able to show that many other nebulæ, which had been frequently observed and had even been sketched, possessed further features too faint and delicate to be seen by any human eye, even with the help of the most powerful telescope. These further features were discovered because they came within the ken of the intensely acute perception of the photographic plate. On the plate these features which the camera showed, were added to those which the eye had already perceived, and when these additions were made it was not infrequently found that the nebula assumed the form of a spiral. But the most remarkable circumstance has still to be added. Some of the plates exposed by Dr. Roberts show clear and unmistakable photographs of spiral nebulæ as exquisite in detail as the Great Spiral itself, but yet so faint that they have never been seen by the eye in any telescope whatever, though they could not elude the photographic plate. Thus, Dr. Roberts not only confirmed in the most splendid manner that really great discovery of the spiral nebulæ of which the honour belongs to Lord Rosse, but the eminent photographic astronomer added many other spirals of the greatest interest to the list of those objects which Lord Rosse had himself given.

Though these discoveries placed the fact of the existence of spiral nebulæ in an impregnable position, and though they greatly increased the interest with which astronomers study such objects, yet another stop had to be taken before the spiral nebula attained the position of extraordinary importance as a celestial object which must now be acknowledged to be its due.

We have already had occasion to mention the marvellous discoveries of nebulæ which the lamented Professor Keeler made with the Crossley Reflector at the Lick Observatory. We have explained that his discoveries have shown the number of nebulæ in the heavens to be probably at least twenty times that which previous observations would have authorised us in asserting. The mere announcement that 120,000 new nebulæ were within the reach of a photographic plate attached to the Crossley Reflector, would, by itself, have been a statement so remarkable as to command the immediate attention of the scientific world. But the interest of even this statement shrinks to unimportance relatively to the further fact which Professor Keeler has added. I do not know, in the annals of astronomy, a pronouncement of greater interest, certainly none of more importance for our present purpose, than the statement that of the 120,000 new nebulæ, at least half are spirals. Here is indeed a stupendous revolution in our knowledge of the celestial objects. Fifty years ago Lord Rosse announced the discovery of a spiral nebula, and the existence of this spiral was doubted at first, though it was gradually conceded at last. Now we have the announcement, on the unchallenged evidence of the photographic plate itself, that to all appearances there are at least 60,000 spiral nebulæ in the heavens. It is, alas! too true that Professor Keeler did not live long enough to enumerate all those nebulæ himself, and, indeed, they have not so far been actually counted, but to those who will study Professor Keeler's papers, the evidence of the substantial accuracy of the statement is incontestable.

And astonishing as this statement may be, we have still to add that, in face of the actual facts, it may be regarded as even a moderate estimate of the abundance of spirals in the universe. We must remember that a spiral nebula is a flat object with long arms extending from it which lie nearly in the same plane. If we are actually to see that such an object is spiral, it is necessary for it to be turned squarely towards the earth. If the object be too much foreshortened, it is quite plain that we can hardly expect to detect its spiral character.

Figure 34. A Ray Nebula (Lord Rosse). (3628 in n.g.c.)
(From the Scientific Transactions of the Royal Dublin Society.)

It is also obvious if the spiral happens to be turned edgeways towards us, that then its spiral form cannot be seen; it would merely appear as what astronomers often call a ray. In the enumeration of the spirals it is therefore only possible for us to include those which happen to be so far squarely turned towards the earth as to make their spiral character unmistakable. We might, therefore, reasonably expect that the numbers of spiral nebulæ actually counted would fall short of the reality. We know that there are many nebulæ of a somewhat elliptical shape. There are also many nebulæ that look like long rays. Those who are familiar with the appearance of nebulæ in great telescopes will recall at once the numerous spindle-shaped objects of this class. It can hardly be doubted that many of the nebulæ, more or less oval in form, and also these rays or the spindle-shaped objects so frequently seen in good telescopes are in reality spiral nebulæ, which are turned not squarely towards us, but which we are merely looking at more or less edgewise, so that they have been foreshortened enough to hide their peculiar structure. Taking these considerations into account, it becomes obvious that the estimate of Professor Keeler as to the number of spiral nebulæ in the heavens, vast as that estimate seems, may still fall short of the truth. Thus we are led to one of the most remarkable conclusions of modern astronomy, viz. that the spiral nebula, next to a star itself, is the most characteristic object in the sidereal heavens.

In treating of the nebulæ in Chapter 4 we explained those fundamental features of the different spectra which make it possible to discriminate with confidence between a nebula which is purely gaseous and a nebula which cannot be so described. As the spiral nebulæ form a class characterised among all the other nebulæ by the possession of a very particular structure, it is interesting to enquire what evidence the spectrum gives with regard to the nature of the material which enters into the constitution of the nebulæ which belong to this strongly-marked group. I do not mean to say that all the 60,000 spirals have been examined with the spectroscope, but, as already explained previously, a sufficient number have been examined to decide the question. We learn from Professor Scheiner, a well-known authority on astronomical spectroscopy, that the spectra of spirals are generally found to be continuous; in other words, we learn that a spiral nebula is not gaseous. It does not consist, like, for example, the nebula in Orion, of vaporous matter in a state of incandescence.

A nebula or a nebulous-looking object which does not give a spectrum of bright lines, but which does give a continuous spectrum, is not infrequently set down as being merely a cluster of stars. This is undoubtedly a true statement with regard to some of these nebulous objects, but it is not true with regard to all. It is much more reasonable to suppose that the greater part of the materials of the spiral nebulæ, though certainly not in the form of gas, are still not condensed into objects large enough to entitle them to be called stars. It must be remembered that when an object of a gaseous nature has lost heat by radiation, and has begun to draw itself together, the gas condenses into particles which constitute small portions of liquid or solid, just as the vapour of water in the atmosphere condenses into the beads of water that form the clouds in our own sky. These small objects, even if incandescent, would no longer radiate light with the characteristics of a gaseous nebula. The light they would emit would be of the same character as that dispensed from the particles of carbon in the solar photosphere to which the sun owes its light. Radiation from such a source would give light with a continuous spectrum, like that from the sun or a star.

From the fact that the spectra of the spiral nebulæ are continuous, we may infer that, though these nebulæ have reached an advanced stage in their development, they have not always, and, perhaps, not generally, attained to the stage in which condensation transformed them into a cluster of actual stars. They have, however, reached a stage in their progress towards those systems of large bodies that they are ultimately to become. The character of its spectrum may show us that the spiral nebula is not very young, that it has attained a considerable age in its evolution as compared with other nebulæ which do not show the spiral character and which have a gaseous spectrum. The importance of this consideration will be made apparent in the next chapter, when we discuss the dynamical conditions to which a spiral nebula must submit.

But there is no reason to doubt that some of the spiral nebulæ may be in reality star-clusters, in which there are aggregations of myriads of points, each justly entitled by its dimensions and its lustre to be regarded as a real star. The great nebula in Andromeda seems to be a greatly foreshortened spiral. This, at least, is the interpretation which may perhaps be most reasonably given to Dr. Roberts' famous photograph of this splendid object. The spectrum of the Andromeda nebula has been photographed by Scheiner after a protracted exposure of seven and a half hours. That spectrum showed no trace of bright lines, thus proving that there is no discernible incandescent gas in the nebula of Andromeda. It gives practically a continuous spectrum, across which some broad bands can be recognised. It was interesting to compare this spectrum of the great nebula in Andromeda with the solar spectrum seen by the same apparatus and under the same conditions. Professor Scheiner announces that there was a remarkable coincidence between the two, and he draws the inference that the stars which enter into the nebula in Andromeda are stars of that particular type to which the sun belongs.

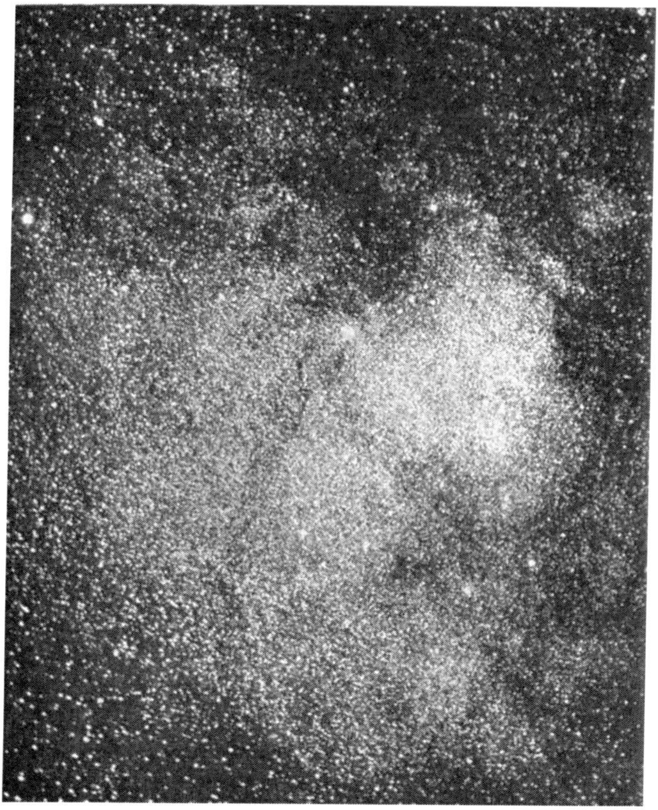

Figure 35. Portion of the Milky Way (near Messier II.). (Photographed by Professor E. E. Barnard.) *(From the Royal Astronomical Society Series.)*

But we have now to point out how the recent study of nebulæ has afforded a yet more striking confirmation of the nebular theory. Laplace showed how a gradually condensing nebula might have formed a sun and a system of planets. Had Laplace known of the spiral nebulæ he would, I doubt not, have found in them the most striking illustration of the operation of evolution on a gigantic scale. They would have provided him with admirable arguments in support of the nebular theory. It is possible that they might also have provided suggestions as to the details of the evolution, which he had not anticipated. But Laplace did not know of such objects, and we can only deplore the loss of the instructive lessons which his incomparable genius would have derived from them.

We must, however, admit that the lessons as to the origin of the solar system, derived from the spiral nebulæ, must be received with due limitation. We may say at once that the *great* spiral nebulæ do not appear to be evolving into systems like the sun and planets; their work is of a higher order of magnitude altogether. The great spiral nebulæ seem to be more analogous to galaxies, like the Milky Way, than to solar systems. The spiral nebula instead of being described as a system, should perhaps be described as a system of systems. If the solar system were drawn to scale on the photograph of the Great Spiral the orbit of Neptune would not be larger than the smallest recognisable dot.

Chapter 11

THE UNERRING GUIDE

We have to consider in this chapter the light which the laws of mathematics throw upon certain features which are possessed by a very large number of celestial objects. Let us first describe, as clearly as the circumstances will permit, the nature of these common features to which we now refer, and of which mathematics will suggest the explanation.

We shall begin with our solar system, in which the earth describes an orbit around the sun. That orbit is contained within a plane, which plane passes through the centre of the sun. We may neglect for the present the earth's occasional slight deviations from this plane which are caused by the attractions of the other planets. If we consider the other bodies of our system, such, for instance, as Venus or Jupiter, we find that the orbit of Venus also lies in a plane, and that plane also passes through the centre of the sun. The orbit of Jupiter is found to be contained within a plane, and it, too, passes through the sun's centre. Each of the remaining planets in like manner is found to revolve in an orbit which is contained in a plane, and all these planes have one common point, that point being the centre of the sun.

It is a remarkable fact that the mutual inclinations are very small, so that the several planes are nearly coincident. If we take the plane of our earth's orbit, which we call the ecliptic, as the standard, then the greatest inclination of the orbit of any other important planet is seven degrees, which is found in the case of Mercury. The inclinations to the ecliptic of the planes of the orbits of a few of the asteroids are much more considerable; to take an extreme case, the orbit of Pallas is inclined at an angle of no less than thirty-four degrees. It must, however, be remembered that the asteroids are very small objects, as the collective masses of the five hundred which are at present known would amount to no more than an unimportant fraction of the mass of one of the great planets of our system. Three-fourths of the asteroids have inclinations under ten degrees. We may, therefore, leave these bodies out of consideration for the present, though we may find occasion to refer to them again later on. Still less need we pay attention at present to the comets, for though these bodies belong to our system, and though they move in plane orbits, which like the orbits of the planets pass through the centre of the sun, yet their orbits are inclined at angles of very varying magnitudes. Indeed, we cannot detect any tendency in the orbits of comets to approximate to the plane of the ecliptic. The masses of comets are, however, inconsiderable in comparison with the robust globes which form the planets, while the origin of comets has been apparently so different from that of the planets, that we may leave them out of consideration in our present argument. There is nothing in the motion of either asteroids or comets to invalidate the general proposition which affirms, that the planes of the orbits of the heaviest and most important bodies in the solar system are very nearly coincident.

Many of the planets are accompanied by satellites, and these satellites revolve round the planets, just as the planet accompanied by its satellites revolves round the sun. The orbit of each satellite is contained within a plane, and that plane passes through the centre of the planet to which it is appended. We thus have a system of planes appropriate to the satellites, just as there is a system of planes appropriate to the planets. The orbits of the satellites of each planet are very nearly in the same plane, with notable exceptions in the cases of Uranus and Neptune, which it will be necessary to consider at full length later on. This plane is very nearly coincident with the planes in which the planets themselves move. Omitting the exceptions, which are unimportant as to magnitude, though otherwise extremely interesting and instructive, the fundamental characteristic of the movements of the principal bodies in our system is that their orbits are nearly parallel to the same plane. We draw an average plane through these closely adjacent planes and we term it the principal plane of our system. It is not, indeed, coincident with the plane of the orbit of any one planet, yet the actual plane of the orbit of every important planet, and of the important satellites, lies exceedingly close to this principal plane. This is a noteworthy circumstance in the arrangement of the planetary system, and we expect that it must admit of some physical explanation.

When we look into the details of the planetary groups composing the solar system, we find striking indications of the tendency of the orbits of the bodies in each subordinate system to become adjusted to a plane. The most striking instance is that exhibited by the Rings of Saturn. It has been demonstrated that these wonderful rings are composed of myriads of separate particles. Each of these particles follows an independent orbit round Saturn. Each such orbit is contained in a plane, and all these planes appear, so far as our observations go, to be absolutely coincident. It is further to be noted that the plane, thus remarkably related to the system of rings revolving around Saturn, is substantially identical with the plane in which the satellites of Saturn themselves revolve, and this plane again is inclined at an angle no greater than twenty-eight degrees to the plane of the ecliptic, and close to that in which Saturn itself revolves around the sun.

Overlooking, as we may for the present, the varieties in detail which such natural phenomena present, we may say that the most noticeable characteristic of the revolutions in the solar system is expressed by the statement that they lie approximately in the same plane.

We shall also find that this tendency of the movements in a system to range themselves in orbits which lie in the same plane, is exhibited in other parts of the universe. Let us consider from this point of view the spiral nebulæ, those remarkable objects which, in the last chapter, we have seen to be so numerous and so characteristic. It is obvious that a spiral nebula must be a flat object. Its thickness is small in comparison with its diameter.

When a spiral nebula is looked at edgewise, then it seems long and thin, so much so that it presents the appearance of a ray such as we have shown in above, which represents a type of object very familiar to those astronomers who are acquainted with nebulæ.

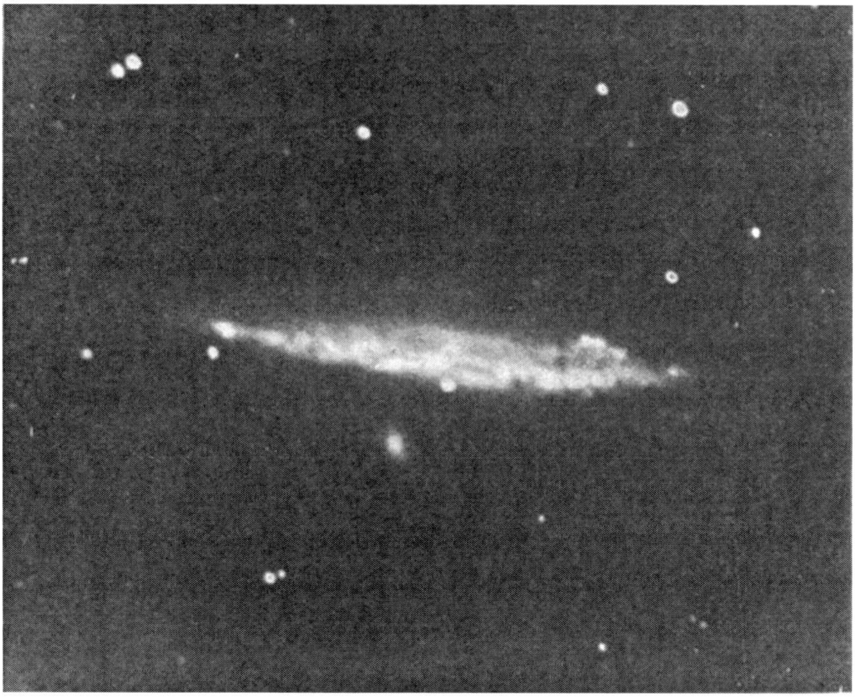

Figure 36. A Spiral Nebula Seen Edgewise (n.g.c. 3628; in Leo). (*Photographed by Dr. Isaac Roberts, F.R.S.*)

The characteristics of these objects seem consistent only with the supposition that there is a tendency in the materials which enter into a spiral nebula to adapt their movements to a particular plane, just as there is a tendency for the objects in Saturn's ring to remain in a particular plane, and just as there has been a tendency among the bodies belonging to the solar system themselves to revolve in a particular plane. Remembering also that there seems excellent reason to believe that spiral nebulæ exhibiting this characteristic are to be reckoned in scores of thousands, it is evident that the fundamental feature in which they all agree must be one of very great importance in the universe.

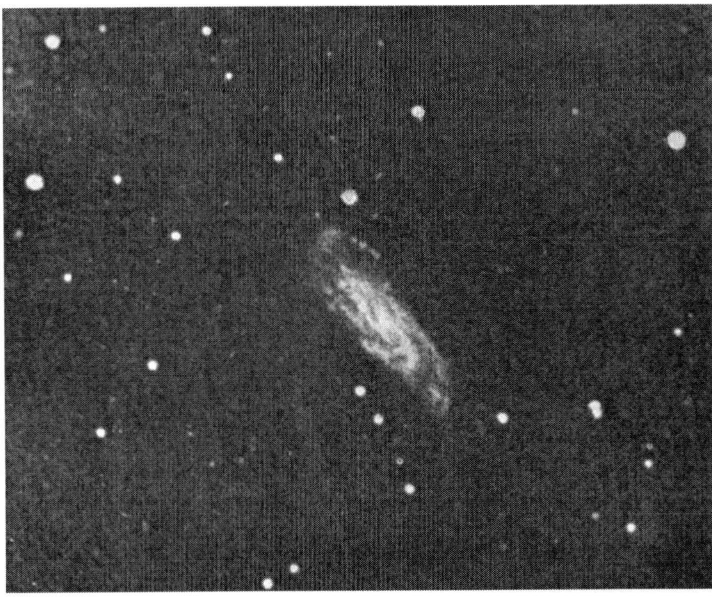

Figure 37. A foreshortened Spiral (n.g.c. 3198; in Ursa Major).
(Photographed by Dr. Isaac Roberts, F.R.S.)

Figure 38. Edge-View of a Spiral boldly shown (n.g.c. 4565; in Coma Berenices).
(Photographed by Dr. Isaac Roberts, F.R.S.)

We may mention yet one more illustration of the remarkable tendency, so frequently exhibited by an organised system in space, to place its parts ultimately in or near the same plane, or at all events, to assume a shape of which one dimension is small in comparison with the two others. We have, in the last chapter, referred to the Milky Way, and we have alluded to the significance of the obvious fact that, however the mass of stars which form the Milky Way may be arranged, they are so disposed that the thickness of the mass is certainly much less than its two other dimensions. Herschel's famous illustration of a grindstone to represent the shape of the Milky Way will serve to illustrate the form we are now considering.

When we meet with a characteristic form so widely diffused through the universe, exhibited not only in the systems attending on the single planets, not only in the systems of planets which revolve round a single sun, but also in that marvellous aggregation of innumerable suns which we find in the Milky Way, and in scores of thousands of nebulæ in all directions, at all distances, and apparently of every grade of importance, we are tempted to ask whether there may not be some physical explanation of a characteristic so universal and so remarkable.

Let us see whether mathematics can provide any suggestion as to the cause of this tendency towards flatness which seems to affect those systems in the universe which are sufficiently isolated to escape from any large disturbance of their parts by outside interference. We must begin by putting, as it were, the problem into shape, and by enumerating certain conditions which, though they may not be absolutely fulfilled in nature, are often so very nearly fulfilled that we make no appreciable error by supposing them to be so.

Let us suppose that a myriad bodies of various sizes, shapes, materials and masses, are launched in space in any order whatever, at any distances from each other, and that they are started with very different movements. Some may be going very fast, some going slowly, or not at all; some may be moving up or down or to the right or to the left—there may be, in fact, every variety in their distances and their velocities, and in the directions in which they are started.

We assume that each pair of masses attract each other by the well-known law of gravitation, which expresses that the force between any two bodies is proportional directly to the product of their masses and inversely to the square of their distance. We have one further supposition to make, and it is an important one. We shall assume that though each one of the bodies which we are considering is affecting all the others, and is in turn affected by them, yet that they are subjected to no appreciable disturbing influence from other bodies not included in the system to which they belong. This may seem at first to make the problem we are about to consider a purely imaginary one, such as could only be applicable to systems different from those which are actually presented to us in nature. It must be admitted that the condition we have inferred can only be approximately fulfilled. But a little consideration will show that the supposition is not an unreasonable one. Take, for instance, the solar system, consisting of the sun, the planets, and their satellites. Every one of these bodies attracts every other body, and the movement of each of the bodies is produced by the joint effects of the forces exerted upon it by all the others. Assuredly this gives a problem quite difficult enough for all the resources that are at our command. But in such investigations we omit altogether the influence of the stars. Sirius, for example, does exercise some attraction on the bodies of our system, but owing to its enormous distance, in comparison with the distances in our solar system, the effect of the disturbance of Sirius on the relative movements of the planets is wholly inappreciable. Indeed, we may add that the disturbances in the solar system produced by all the stars, even including the myriads of the Milky Way, are absolutely negligible. The movements in our solar system, so far as our observations reveal them, are performed precisely as if all bodies of the universe foreign to the solar system were non-existent. This consideration shows that in the problem we are now to consider, we are introducing no unreasonable element when we premise that the system whose movements we are to investigate is to be regarded as free from appreciable disturbance by any foreign influence.

To follow the fortunes of a system of bodies, large or small, starting under any arbitrary conditions at the commencement, and then abandoned to their mutual attractions, is a problem for the mathematician. It certainly presents to him questions of very great difficulty, and many of these he has to confess are insoluble; there are, however, certain important laws which must be obeyed in all the vicissitudes of the motion. There are certain theorems known to the mathematician which apply to such a system, and it is these theorems which afford us most interesting and instructive information. I am well aware that the subject upon which I am about to enter is not a very easy one, but its importance is such that I must make the effort to explain it.

Let me commence by describing what is meant when we speak of the energy of a system. Take, first, the case of merely two bodies, and let us suppose that they were initially at rest. The energy of a system of this very simple type is represented by the quantity of work which could be done by allowing these two bodies to come together. If, instead of being in the beginning simply at rest, the bodies had each been in motion, the energy of the system would be correspondingly greater. The energy of a moving body, or its capacity of doing work in virtue of its movement, is proportional jointly to its mass and to the square of its velocity. The energy of the two moving bodies will therefore be represented by three parts; first, there will be that due to their distance apart; secondly, there will be that due to the velocity of one of them; and, thirdly, there is that due to the velocity of the other. In the case of a number of bodies, the energy will consist in the first place of a part which is due to the separation of the bodies, and measured by the quantity of work that would be produced if, in obedience to their mutual attraction, all the bodies were allowed to come together into one mass. In the second place, the bodies are to be supposed to have been originally started with certain velocities, and the energy of each of the bodies, in virtue of its motion, is to be measured by the product of one-half its mass into the square of its velocity. The total energy of the system consists, therefore, of the sum of the parts due to the velocities of the bodies, and that which is due to their mutual separation.

If the bodies could really be perfectly rigid, unyielding masses, so that they have no movements analogous to tides, and if their movements be such that collisions will not take place among them, then the laws of mechanics tell us that the quantity of energy in that system will remain forever unaltered. The velocities of the particles may vary, and the mutual distances of the particles may vary, but those variations will be always conducted, subject to the fundamental condition that if we multiply the square of the velocity of each body by one-half its mass, and add all those quantities together, and if we increase the sum thus obtained by the quantity of energy equivalent to the separation of the particles, the total amount thus obtained is constant. This is the fundamental law of mechanics known as the conservation of energy.

For such material systems as the universe presents to us, the conservation of energy, in the sense in which I have here expressed it, will not be maintained; for the necessary conditions cannot be fulfilled. Let us suppose that the incessant movements of the bodies in the system, rushing about under the influence of their mutual attractions, has at last been productive of a collision between two of the bodies. We have already explained in Chapter 6. how in the collision of two masses the energy which they possess in virtue of their movements may be to a large extent transformed into heat; there is consequently an immediate increase in the temperature of the bodies concerned, and then follows the operation of that fundamental law of heat, by which the excess of heat so arising will be radiated away. Some of it will, no doubt, be intercepted by falling on other bodies in the system, and the amount that might be thus possibly retained would, of course, not be lost to the system. The bodies of the solar system at least are so widely scattered, that the greater part of the heat would certainly escape into space, and the corresponding quantity of energy would be totally lost to the system. We may generally assume that a collision among the bodies would be most certainly productive of a loss of energy from the system.

No doubt collisions can hardly be expected to occur in a system consisting of large, isolated bodies like the planets. Even in any system of solid bodies collisions may be presumed to be infrequent in comparison with the numbers of the bodies. But if, instead of a system of few bodies of large mass, we have a gas or nebula composed of innumerable atoms or molecules, the collisions would be by no means infrequent, and every collision, in so far as it led to the production of heat, would be productive of loss of energy by radiation from the system.

It should also be added that, even independently of actual collisions, there is, and must be, loss of energy in the system from other causes. There are no absolutely rigid bodies known in nature, for the hardest mineral or the toughest steel must yield to some extent when large forces are applied to it, and as the bodies in the system are not mere points or particles of inconsiderable dimensions, they will experience stresses something like those to which our earth is subjected in that action of the moon and sun which produces the tides. In consequence of the influences of each body on the rest, there will be certain relative changes in the parts of each body; there will be, as it were, tidal movements in their liquid parts and even in their solid substance. These tides will produce friction, and this will produce heat. This heat will be radiated from the system, but the heat radiated corresponds to a certain amount of energy; the energy is therefore lost to the system, so that even without actual collisions we still find that energy must be gradually lost to the system.

Thus we have been conducted to an important conclusion, which may be stated in the following way. Let there be any system of bodies, subject to their mutual attractions, and sufficiently isolated from the disturbing influence of all bodies which do not belong to the system, then the original energy with which that system is started must be undergoing a continual decline. It must at least decline until such a condition of the system has been reached that collisions are no longer possible and that tidal influences have ceased. These conditions might be fulfilled if all the bodies of the system coalesced into a single mass.

As illustrations of the systems we are now considering, we may take the sun and planets as a whole. A spiral nebula is a system in the present sense, while perhaps the grandest illustration of all is provided by the Milky Way.

It will be noted that we may have a system which is isolated so far as our present argument is concerned, even while it forms a part of another system of a higher order of magnitude. For instance, Saturn with his rings and satellites is sufficiently isolated from the rest of the solar system and the rest of the universe, to enable us to trace the consequences of the gradual decline of energy in his attendant system. The solar system in which Saturn appears merely as a unit, is itself sufficiently isolated from the stars in the Milky Way to permit us to study the decline of energy in the solar system, without considering the action of those stars.

This general law of the decline of energy in an isolated system, is supplemented by another law often known as the conservation of moment of momentum. It may at first seem difficult to grasp the notion which this law involves. The effort is, however, worth making, for the law in question is of fundamental importance in the study of the mechanics of the universe. In the Appendix will be found an investigation by elementary geometry of the important mechanical principles which are involved in this subject.

Whatever may have been the origin of the primæval nebula, and whatever may have been the forces concerned in its production we may feel confident that it was not originally at rest. We do not indeed know any object which is at rest. Not one of the heavenly bodies is at rest, nothing on earth is at rest, for even the molecules of rigid matter are in rapid motion. Rest seems unknown in the universe. It would be, therefore, infinitely improbable that a primæval nebula, whatever may have been the agency by which it was started on that career which we are considering, was initially in a condition of absolute rest. We assume without hesitation that the nebula was to some extent in motion, and we may feel assured that the motions were of a highly complicated description. It is fortunate for us that our argument does not require us to know the precise character of the movements, as such knowledge would obviously be quite unattainable. We can, however, invoke the laws of mechanics as an unerring guide. They will tell us not indeed everything about those motions, but they will set forth certain characteristics which the movements must have had, and these characteristics suffice for our argument.

To illustrate the important principle on which we are now entering I must mention the famous problem of three bodies which has engaged the attention of the greatest mathematicians. Let there be a body A, and another B, and another C. We shall suppose that these bodies are so small that they may be regarded merely as points in comparison with the distances by which they are separated. We shall suppose that they are all moving in the same plane, and we shall suppose that each of them attracts the others, but that except these attractions there are no other forces in the system. To discover all about the motions of these bodies is so difficult a problem that mathematicians have never been able to solve it. But though we are not able to solve the problem completely, we can learn something with regard to it.

We represent by arrows in Figure 39 the directions in which A, B, and C are moving at the moment. We choose any point O in the plane, and for simplicity we have so drawn the figure that A, B, and C are forces tending to turn round O in the same direction. The velocity of a body multiplied into its mass is termed the *momentum* of the body. Draw the perpendicular from O to the direction in which the body A is moving, then the product of this perpendicular and the momentum of A is called the *moment of momentum* of A around O. In like manner we form the moment of momentum of B and C, and if we add them together we obtain the total moment of momentum of the system.

We can now give expression to a great discovery which mathematicians have made. No matter how complicated may be the movements of A, B, and C; no matter to what extent these particles approximate or how widely they separate; no matter what changes may occur in their velocities, or even what actual collision may take place, the sum of the moments of momentum must remain forever unaltered. This most important principle in dynamics is known as the conservation of moment of momentum.

Though I have only mentioned three particles, yet the same principle will be true for any number. If it should happen that any of them are turning round O in the opposite direction, then their moments of momentum are to be taken as negative. In this case we add the moments tending in one direction together; and then subtract all the opposite moments. The remainder is the quantity which remains constant.

We may state this principle in a somewhat different manner as follows: Let us consider a multitude of particles in a plane; let them be severally started in any directions in the plane, and then be abandoned to their mutual attractions, it being understood that there are no forces produced by bodies external to the system; if we then choose any point in the plane, and measure the areas described round that point by the several moving bodies in one second, and if we multiply each of, those areas by the mass of the corresponding body, then, if all the bodies are moving in the same direction round the point, the sum of the quantities so obtained is constant. It will be the same a hundred or a thousand years hence as it is at the present moment, or as it was a hundred or a thousand years ago. If any of the particles had been turning round the point in the opposite direction, then the products belonging to such particles are to be subtracted from the others instead of added.

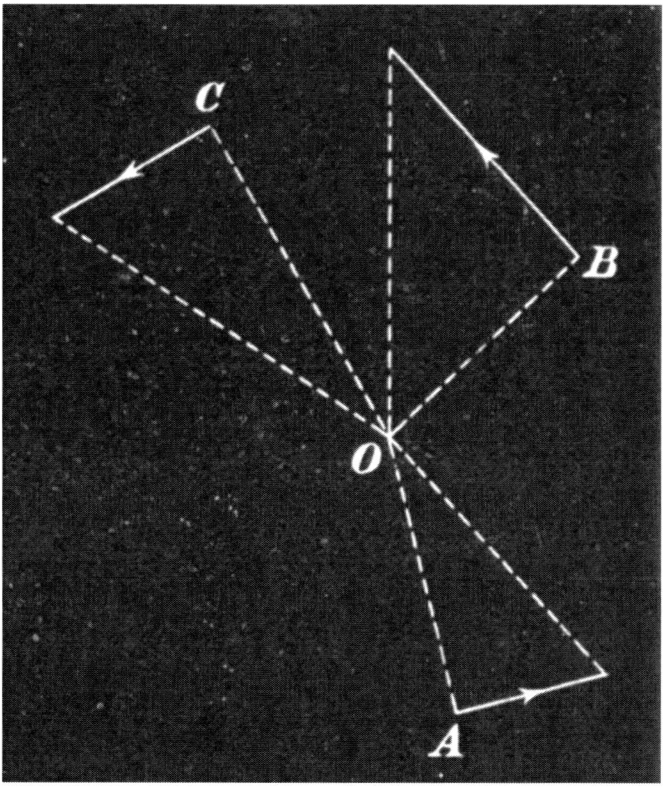

Figure 39. To illustrate Moment of Momentum.

We have now to express in a still more general manner the important principle that is here involved. Let us consider any system of attracting particles, no matter what their masses or whether their movements be restricted to a plane or not. Let us start them into motion in any directions and with any initial velocities, and then abandon them to the influence of their mutual attractions, withholding at the same time the interference of any forces from bodies exterior to the system. Draw any plane whatever, and let fall perpendiculars upon this plane from the different particles of the system. It will be obvious that as the particles move the feet of the perpendiculars must move in correspondence with the particles from which the perpendiculars were let fall. We may regard the foot of every perpendicular as the actual position of a moving point, and it can be proved that if the mass of each particle be multiplied into the area which the foot of its perpendicular describes in a second round any point in the plane, and then be added to the similar products from all the other particles, only observing the proper precautions as to sign, the sum will remain constant, i.e., in any other second the total quantity arrived at will be exactly the same. This is a general law of dynamics. It is not a law of merely approximate truth, it is a law true with absolute accuracy during unlimited periods of time.

The actual value of the constant will depend both on the system and on the plane. For a given system the constant will differ for the different planes which may be drawn, and there will be some planes in which that sum will be zero. In other words, in those planes the areas described by the feet of the perpendiculars, multiplied by the masses of the particles which are moving in one way, will be precisely equal to the similar sum obtained from the particles moving in the opposite direction.

But among all possible planes there is one of special significance in its relation to the system. It is called the "principal plane," and it is characterised by the fact that the sum (with due attention to sign) of the areas described each second by the feet of the perpendiculars, multiplied into the masses of the corresponding particle, is greater than the like magnitude for any other plane, and is thus a maximum. For all planes parallel to this principal plane, the result will be, of course, the same; it is the direction of the plane and not its absolute situation that is material. We thus see that while this remarkable quantity is constant in any plane, for all time, yet the actual value of that constant depends upon the aspect of the plane; for some planes it is zero, for others the constant has intermediate values, and there is one plane for which the constant is a maximum. This is the principal plane, and a knowledge of it is of vital importance in endeavouring to understand the nebular theory. Nor are the principles under consideration limited only to a system consisting of sun and planets; they apply, with suitable modifications, to many other celestial systems as well.

The instructive character of this dynamical principle will be seen when we deduce its consequences. The term "moment of momentum" of a particle, with reference to a certain point in a plane, expresses double the product of the rate at which the area is described by the foot of the perpendicular to this plane, multiplied by the mass of the particle. The moment of momentum of the system, with reference to the principal plane, is a maximum in comparison with all other planes; that moment of momentum retains precisely the same value throughout all time, from the first instant the system was started onwards. And it retains this value, no matter what changes or disturbances may happen in the system, provided only that the influence of external forces is withheld. Subject to this condition, the transformations of the system may be any whatever. The several bodies may be forced into wide changes of their orbits, so that there may even be collisions among them; yet, notwithstanding those collisions, and notwithstanding the violent alterations which may be thus produced in the movements of the bodies, the moment of momentum will not alter. No matter what tides may be produced, even if those tides be so great as to produce disruption in the masses and force the orbits to change their character radically, yet the moment of momentum will be conserved without alteration.

It is essential to notice the fundamental difference between the principle which has been called the conservation of energy in the system, and the conservation of moment of momentum. We have pointed out that when collisions take place, part of the energy due to motion is transformed into heat, and energy in that form admits of radiation through space, and thus becomes lost to the system, with the result that the total energy declines. Even without actual collision, we have shown how certain effects of tides, or other consequences of friction, necessarily involve the squandering of energy with which the system was originally endowed. A system started with a certain endowment of energy may conserve that energy indefinitely, if all such actions as collisions or frictions are absent. If collisions or frictions are present the system will gradually dissipate energy. Our interpretation of the future of such a system must always take account of this fundamental fact.

It is, of course, conceivable that the moment of momentum with which a system was originally endowed might have happened to be zero. A system of particles could be so constructed and so started on their movements that their moment of momentum with regard to a certain plane should be zero. It might happen that the moment of momentum of the system with regard to a second plane, perpendicular to the former one, should be also zero; and, finally, that the moment of momentum of the system with regard to a third plane perpendicular to each of the other two, should be also zero. If these three conditions were found to prevail at the commencement, they would prevail throughout the movement, and, more generally still, we may state that in such circumstances the moment of momentum of the system would be zero about any plane whatever. There would be no principal plane in such a system. We thus note that though it is inconceivable that a group of mutually attracting bodies should be started into movement without a suitable endowment of energy, it is yet quite conceivable that a system could be started without having any moment of momentum. And if at the beginning the system had no moment of momentum, then no matter what may be the future vicissitudes of its motion, no moment of momentum can ever be acquired by it to all eternity, so long as the interference of external forces is excluded.

But having said this much as to the conceivability of the initiation of a system with no moment of momentum, we now hasten to add that, so far as Nature is actually concerned, this bare possibility may be set aside as one which is infinitely improbable. Nature does not do things which are infinitely improbable, and, therefore, we may affirm that all material systems, with which we shall have to deal, do possess moment of momentum. However the system may have originated, whatever may have been the actions of forces by which it was brought into being, we may feel assured that the system received at its initiation some endowment of moment of momentum, as well as of energy. Hence we may conclude that every such system as is presented to us in the infinite variety of Nature, must stand in intimate relation to some particular plane, being that which is known as the principal plane of moment of momentum. In our effort to interpret Nature, the physical importance of this fact can hardly be over-estimated.

In a future chapter we shall make some attempt to sketch the natural operations by which individual systems have been started on their careers. Postponing, then, such questions, we propose to deal now with the phenomena which the principles of dynamics declare must accompany the evolution of a system under the action of the exclusive attraction of the various parts of that system for each other. The system commences its career with a certain endowment of energy, with a certain endowment of moment of momentum, and with a certain principal plane to which that moment of momentum is specially related. In the course of the evolution through which, in myriads of ages, the system is destined to pass, the energy that it contains will undergo vast loss by dissipation. On the other hand, the moment of momentum will never vary, and the position of the principal plane will remain the same for all time. We have to consider what features, connected with the evolution, may be attributed to the operation of these dynamical laws. We have, in fact, to deduce the consequences which seem to follow from the fact that, in consequence of collisions, and in consequence of friction, an isolated system in space must gradually part with its initial store of energy, but that, notwithstanding any collisions and any friction, the total moment of momentum of the system suffers no abatement.

As the system advances in development, we have to deal with a gradual decline in the ratio of the original store of energy to the original store of moment of momentum. And hence we must expect that a system will ultimately tend towards a form in which, while preserving its moment of momentum, it shall do so with such a distribution of the bodies of which it consists as shall be compatible with a diminishing quantity of energy. It is not hard to see that in the course of ages this tends, as one consequence, to make the movements of each of the bodies in the system ultimately approximate to movements in a plane.

Let us, for simplicity, begin with the case of three attracting particles, A, B and C. Let B be started in any direction in the plane L, and let A be started in an orbit round it, and in the same plane L. Now let C be started into motion, in any direction, from some point also in L. It is certain that the sum of the areas projected parallel to any plane, which are described in a second by these three bodies, must be constant, each of the areas being, as usual, multiplied by the mass of the corresponding body. Let us specially consider the plane L in which the motions of A and B already lie. It is on this plane that the area described by C has to be projected. The essential point now to remember is that the projected area is less than the actual area. It is plain that if C has to describe a certain projected area in a certain time, the velocity with which C has to move must be greater when C starts off at an inclination to the plane than would have been necessary if C had started in the plane, other things being the same. Thus we see that, if the three bodies were all moving in the same plane, they could, speaking generally, maintain more easily the requisite description of areas, that is, the requisite moment of momentum with smaller velocities than if they were moving in directions which were not so regulated; that is to say, the moment of momentum can be kept up with less energy when the particles move in the same plane.

In a more general manner we see that any system in which the bodies are moving in the same plane will, for equal moment of momentum, require less energy than it would have done had the bodies been moving in directions which were not limited to a plane. Thus we are led to the conclusion that the ultimate result of the collisions and the friction and the tides, which are caused by the action of one particle on another, is to make the movements tend towards the same plane.

In this dynamical principle we have in all probability a physical explanation of that remarkable characteristic of celestial movements to which we have referred. The solar system possesses less energy in proportion to its moment of momentum than it would require to have if the orbits of the important planets, instead of lying practically in the same plane, were inclined at various angles. Whatever may have been the original disposition of the materials forming the solar system, they must once have contained much more energy than they have at present. The moment of momentum in the principal plane, at the beginning, was not, however, different from the moment of momentum that the system now possesses. As the energy of the system gradually declined, the system has gradually been compelled to adjust itself in such a manner that, with the reduced quantity of energy, the requisite moment of momentum shall still be preserved. This is the reason why, in the course of the myriads of ages during which the solar system has been acquiring its present form, the movements have gradually become nearly conformed to a plane.

The operation of the principle, now before us, may be seen in a striking manner in Saturn's ring. The particles constituting this exquisite object, so far as observations have revealed them, seem to present to us an almost absolutely plane movement. The fact that the movements of the constituents of Saturn's ring lie in a plane is doubtless to be accounted for by the operation of the fundamental dynamical principle to which we have referred. Saturn, in its great motion round the luminary, is, of course, controlled by the sun, yet the system attached to Saturn is so close to that globe as to be attracted by the sun in a manner which need not here be distinguished from the solar attraction on Saturn itself. It follows that the differential action, so to speak, of the sun on Saturn, and on the myriad objects which constitute its ring, may be disregarded. We are therefore entitled, as already mentioned, to view Saturn and its system as an isolated group, not acted upon by any forces exterior to the system. It is therefore subject to the laws which declare that, though the energy declines, the moment of momentum is to remain unaltered. This it is which has apparently caused the extreme flatness of Saturn's ring. The energy of the rotation of that system has been expended until it might seem that no more energy has been left than just suffices to preserve the unalterable moment of momentum, under the most economical conditions, so far as energy is concerned.

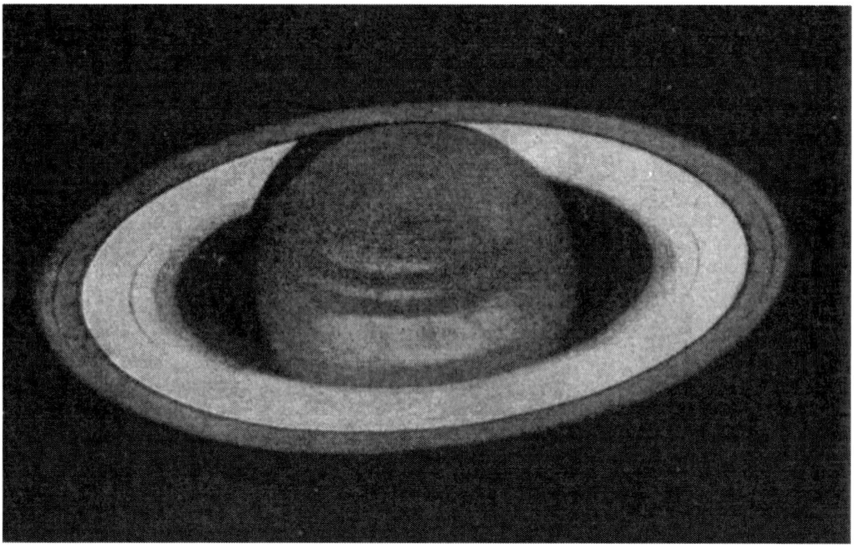

Figure 40. Saturn. Drawn by E. M. Antoniadi. (July 30th, 1899.)

Let us suppose that one of the innumerable myriads of particles which constitute the ring of Saturn were to forsake the plane in which it now revolves, and move in an orbit inclined to the present plane. We shall suppose that the original track of the orbit was a circle, and we shall assume that in the new plane to which the motion is transferred the motion is also circular. That particle will have still to do its share of preserving the requisite total moment of momentum, for we are to suppose that each of the other particles remains unaltered in its pace and in the other circumstances of its motion. The aberrant particle will describe, in a second, an area which, for the purpose of the present calculation, must be projected upon the plane containing the other particles. The area, when projected, must still be as large as the area that the particle would have described if it had remained in the plane. It is therefore necessary that the area swept over by the particle in the inclined plane, in one second, shall be greater than the area which sufficed in the original plane. This requires the circle in which the particle revolves to be enlarged, and this necessitates that its energy should be increased. In other words, while the moment of momentum was no greater than before, the energy of the system would have to be greater. We thus see that inasmuch as the particles forming the rings of Saturn move in circles in the same plane, they require a smaller amount of energy in the system to preserve the requisite moment of momentum than would be required if they moved in circular orbits which were not in the same plane. In such a system as Saturn's ring, in which the particles are excessively numerous and excessively close together, it may be presumed that there may once have been sufficient collisions and frictions among the particles to cause the exhaustion of energy to the lowest point at which the moment of momentum would be sustained. In the course of ages this has been accomplished by the remarkable adjustment of the movements to that plane in which we now find them.

The importance of this subject is so great that we shall present the matter in a somewhat different manner as follows: We shall simplify the matter by regarding the orbits of the planets or other bodies as circles The fact that these orbits are ellipses, which are, however, very nearly circles, will not appreciably affect the argument.

Let us, then, suppose a single planet revolving round a fixed sun, in the centre. The energy of this system has two parts. There is first the energy due to the velocity of the planet, and this is found by taking half the product of the mass of the planet and the square of its velocity. The second part of the energy depends, as we have already explained, on the distance of the planet from the sun. The planet possesses energy on account of its situation, for the attraction of the sun on the planet is capable of doing work. The further the planet is from the sun the larger is the quantity of energy that it possesses from this cause. On the other hand, the further the planet is from the sun the smaller is its velocity, and the less is the quantity of energy that it possesses of the first kind. We unite the two parts, and we find that the net result may be expressed in the following manner: If a planet be revolving in a circular path round the sun, then the total energy of that system (apart from any rotation of the sun and planet on their axes), when added to the reciprocal of the distance between the two bodies, measured with a proper unit of length, is the same for all distances of the same two bodies. This shows the connection between the energy and the distance of the planet from the sun.

Thus we see that if the circle is enlarged the energy of the system increases. The moment of momentum of the system is proportional to the square root of the distance of the two bodies. If, therefore, the distance of the two bodies is increased, the moment of momentum increases also.

It will illustrate the application of the argument to take a particular case in which a system of particles is revolving round a central sun in circular orbits, all of which lie in the same plane. Let us suppose that, while the moment of momentum of the system of particles is to remain unaltered, one of the particles is to be shifted into a plane which is inclined at an angle of 60° to the plane of the other orbits; it can easily be seen that an area in the new plane, when projected down into the original plane, will be reduced to half its amount. Hence, as the moment of momentum of the whole system is to be kept up, it will be necessary for the particle to have a moment of momentum in the circle which it describes in the new plane which is double that which it had in the original plane. It follows that the radius of the circle in the new plane must be four times the radius of the circle which defined the orbit of the particle in the old plane. The energy of the particle in this orbit is therefore correspondingly greater, and thus the energy of the whole system is increased. This illustrates how a system, in which the circular orbits are in different planes, requires more energy for a given moment of momentum than would suffice if the circular orbits had all been in the same plane. So long as the orbits are in different planes there will still remain a reserve of energy for possible dissipation. But the dissipation is always in progress, and hence there is an incessant tendency towards a flattening of the system by the mutual actions of its parts.

It may help to elucidate this subject to state the matter as follows: The more the system contracts, the faster it must generally revolve; this is the universal law when disturbing influences are excluded. Take, for instance, the sun, which is at this moment contracting on account of its loss of heat. In consequence of that contraction it is essential that the sun shall gradually turn faster round on its axis. At present the sun requires twenty-five days, four hours and twenty-nine minutes for each rotation. That period must certainly be diminishing, although no doubt the rate of diminution is very slow. Indeed, it is too slow for us to observe; nevertheless, some diminution must be in progress. Applying the same principle to the primitive nebula, we see, that as the contraction of the original volume proceeds, the speed with which the several parts will rotate must increase.

The periodic times of the planets are here instructive. The materials now forming Jupiter were situated towards the exterior of the nebula, so that, as the nebula contracted, it tended to leave Jupiter behind. The period in which Jupiter now revolves round the sun may give some notion of the period of the rotation of the nebula at the time that it extended so far as Jupiter. Subsequently to the formation, and the detachment of Jupiter, a body which was henceforth no longer in contact with the nebula, the latter proceeded further in its contraction. Passing over the intermediate stages, we find the nebula contracting until it extended no further than the line now marked by the earth's orbit; the speed with which the nebula was rotating must have been increasing all the time, so that though the nebula required several years to go round when it extended as far as Jupiter, only a fraction of that period was necessary when it had reached the position indicated by the earth's track at the present time. Leaving the earth behind it, just as it had previously left Jupiter, the nebula started on a still further condensation. It drew in, until at last it reached a further stage by contraction into the sun, which rotates in less than a month. Thus the period of Jupiter namely, twelve years, the period of the earth, namely, one year, and the period of the sun, namely, twenty-five days, illustrate the successive accelerations of the rotation of the nebula in the process of contraction. No doubt these statements must be received with much qualification, but they will illustrate the nature of the argument.

We may also here mention the satellites of Uranus, all the more so because it has been frequently urged as an objection to the nebular theory that the orbits of the satellites of Uranus lie in a plane which is inclined at a very large angle; no less than 82° to the general plane of the solar system. I shall refer in a later chapter to this subject, and consider what explanation can be offered with regard to the great inclination of this plane, which is one of the anomalies of our system. For the present I merely draw attention to the fact that the movements of all four satellites of Uranus do actually lie in the same plane, though, as already indicated, it stands nearly at right angles to the ecliptic.

Professor Newcomb has shown that the four satellites of Uranus revolve in orbits which are almost exactly circular, and which, so far as observation shows, are absolutely in the same plane. From our present point of view this is a matter of much interest. Whatever may have been the influence by which this plane departs so widely from the plane of the ecliptic, it seems certain that it must be regarded as having acted at a very early period in the evolution of the Uranian system; and when this system had once started on its course of evolution, the operation of that dynamical principle to which we have so often referred was gradually brought to bear on the orbits of the satellites. We have here another isolated case resembling that of Saturn and its rings. The fundamental law ordained that the moment of momentum of Uranus and its moons must remain constant, though the total quantity of energy in that system should decline. In the course of ages this has led to the adjustment of the orbits of the four satellites into the same plane.

I ought here to mention that the rotation of Uranus on its axis presents a problem which has not yet been solved by telescopic observation. It is extremely interesting to note that, as a rule, the axes on which the important planets rotate are inclined at no great angles to the principal plane of the solar system. The great distance of Uranus has, however, prevented astronomers from studying the rotation of that planet in the ordinary manner, by observation of the displacement of marks on its surface. So far as telescopic observations are concerned, we are therefore in ignorance as to the axis about which Uranus revolves. If, following the analogy of Jupiter, or Saturn, or Mars, or the earth, the rotation of Uranus was conducted about an axis, not greatly inclined from the perpendicular to the ecliptic, then the rotation of Uranus would be about an axis very far from perpendicular to the plane in which its satellites revolve. The analogy of the other planets seems to suggest that the rotation of a planet should be nearly perpendicular to the plane in which its satellites revolve. As the question is one which does not admit of being decided by observation, we may venture to remark that the necessity for a declining ratio of energy to moment of momentum in the Uranian system provides a suggestion. The moment of momentum of a system, such as that of Uranus and its satellites, is derived partly from the movements of the satellites and partly from the rotation of the planet itself. From the illustrations we have already given, it is plain that the requisite moment of momentum is compatible with a comparatively small energy only when the system is so adjusted that the axis of rotation of the planet is perpendicular to the plane in which the satellites revolve, or in other words when the satellites revolve in the plane of the equator of the planet. We do not expect that this condition will be complied with to the fullest extent in any members of the solar system. There is indeed an obvious exception; for the moon, in its revolution about the earth, does not revolve exactly in the earth's equator. We might, however, expect that the tendency would be for the movements to adjust themselves in this manner. It seems therefore likely that the direction of the axis of Uranus is perpendicular, or nearly so, to the plane of the movements of its satellites.

At this point we take occasion to answer an objection which may perhaps be urged against the doctrine of moment of momentum as here applied. I have shown that the tendency of this dynamical principle is to reduce the movements towards one plane. It may be objected that if there is this tendency, why is it that the movements have not all been brought into the same plane exactly? This has been accomplished in the case of the bodies forming Saturn's ring, and perhaps in the satellites of Uranus. But why is it that all the great planets of our solar system have not been brought to revolve absolutely in the same plane?

We answer that the operations of the forces by which this adjustment is effected are necessarily extremely slow. The process is still going on, and it may ultimately reach completion. But it is to be particularly observed that the nearer the approach is made to the final adjustment, the slower must be the process of adjustment, and the less efficient are the forces tending to bring it about. For the purpose of illustrating this, we may estimate the efficiency of the forces in flattening down the system in the following manner. Suppose that there are two circular orbits at right angles to each other, and that we measure the efficiency of the action tending to bring the planes to coincide by 100. When the planes are at an angle of thirty degrees the efficiency is represented by 50, and when the inclination is only five degrees the efficiency is no more than 9, and the efficiency gradually lessens as the angle declines. As the angles of inclination of the planes in the solar system are so small, we see that the efficiency of the flattening operation in the solar system must have dwindled correspondingly. Hence we need not be surprised that the final reduction of the orbits into the same plane has not yet been absolutely completed.

Certainly the most numerous, and perhaps the grandest, illustrations of the operation of the great natural principles we have been considering are to be found in the case of the spiral nebulæ. The characteristic appearance of these objects demands special explanation, and it is to dynamics we must look for that explanation.

As to the original cause of a nebula we shall have something to say in a future chapter. At present we are only considering how, when a nebula has come into existence, the action of known dynamical principles will mould that nebula into form. As an illustration of a nebula, in what we may describe as its comparatively primitive shape, we may take the Great Nebula in Orion. This stupendous mass of vaguely diffused vapour may probably be regarded as in an early stage when contrasted with the spirals. We have already shown how the spectroscopic evidence demonstrates that the famous nebula is actually a gaseous object. It stands thus in marked contrast with many other nebulæ which, by not yielding a gaseous spectrum, seem to inform us that they are objects which have advanced to a further stage in their development than such masses of mere glowing gas as are found in the splendid object in Orion.

The development of a nebula must from dynamical principles proceed along the lines that we have already indicated. We shall assume that the nebula is sufficiently isolated from surrounding objects in space as to be practically free from disturbing influences produced by these objects. We shall therefore suppose that the evolution of the nebula proceeds solely in consequence of the mutual attractions of its various parts. In its original formation the nebula receives a certain endowment of energy and a certain endowment of moment of momentum; the mere fact that we see the nebula, the fact that it radiates light, shows that it must be expending energy, and the decline of the energy will proceed continuously from the formation of the object. The laws of dynamics assure us that no matter what may be the losses of energy which the nebula suffers through radiation or through the collisions of its particles, or through their tidal actions, or in any way whatever from their mutual actions, the moment of momentum must remain unchanged.

As the ages roll by, the nebula must gradually come to dispose itself, so that the moment of momentum shall be maintained, notwithstanding that the energy may have wasted away to no more than a fraction of its original amount. Originally there was, of course, one plane, in which the moment of momentum was a maximum. It is what we have called the principal plane of the system, and the evolution tends in the direction of making the nebula gradually settle down towards this plane. We have seen that the moment of momentum can be sustained with the utmost economy of energy by adjusting the movements of the particles so that they all take place in orbits parallel to this plane, and the mutual attractions of the several parts will gradually tend to bring the planes of the different orbits into coincidence. Every collision between two atoms, every ray of light sent forth, conduce to the final result. Hence it is that the nebula gradually tends to the form of a flat plane. This is the first point to be noticed in the formation of a spiral nebula.

But there is a further consideration. As the nebula radiates its light and its heat, and thus loses its energy, it must be undergoing continual contraction. Concurrently with its gradual assumption of a flat form, the nebula is also becoming smaller. Here again that fundamental conception of the conservation of moment of momentum will give us important information. If the nebula contracts, that is to say, if each of its particles draws in closer to the centre, the orbits of each of its particles will be reduced. But the quantity of areas to be described each second must be kept up. We have pointed out that it is infinitely improbable the system should have been started without any moment of momentum, and this condition of affairs being infinitely improbable, we dismiss any thought of its occurrence. As the particles settle towards the plane, the areas swept out by the movements to the right, and those areas swept out by the movements to the left, will not be identical; there will therefore be a balance on one side, and that balance must be maintained without the slightest alteration throughout all time. As the particles get closer together, and as their orbits lessen, it will necessarily happen that the velocities of the particles must increase, for not otherwise can the fundamental principle of the constant moment of momentum be maintained. And as the system gets smaller and smaller, by contraction from an original widely diffused nebulosity, like, perhaps, the nebula in Orion, down to a spiral nebula which may occupy not a thousandth or a millionth part of the original volume, the areas will be kept up by currents of particles moving in the two opposite ways around a central point. As the contraction proceeds, the opposing particles will occasionally collide, and consequently the tendency will be for the predominant side to assert itself more and more, until at last we may expect a condition to be reached in which all the movements will take place in one direction, and when the sum of the areas described in a second, by each of the particles, multiplied by their respective masses, will represent the original endowment of moment of momentum. Thus we find that the whole object becomes ultimately possessed of a movement of rotation.

The same argument will show that the inner parts of the nebula will revolve more rapidly than those in the exterior. Thus we find the whirlpool structure produced, and thus we obtain an explanation, not only of the flatness of the nebula, but also of the spiral form which it possesses. It is not too much to say that the operation of the causes we have specified, if external influence be withheld, tends ultimately to produce the spiral, whatever may have been the original form of the object. No longer, therefore, need we feel any hesitation in believing the assurance of Professor Keeler that out of the one hundred and twenty thousand nebulæ, at least one-half must be spirals. We have found in dynamics an explanation of that remarkable type of object which we have now reason to think is one of the great fundamental forms of nature.

Chapter 12

THE EVOLUTION OF THE SOLAR SYSTEM

We shall consider in this chapter what we believe to have been the history of that splendid system, formed by the planets under the presiding control of the sun. The ground over which we have already passed will prepare us for the famous doctrine that the sun, the planets and their satellites, together with the other bodies which form the group we call the solar system, have originated from the contraction of a primæval nebula.

As the ages rolled by, this great primæval nebula began to undergo modification. In accordance with the universal law which we find obeyed in our laboratories, and which we have reason to believe must be equally obeyed throughout the whole extent of space, this nebula, if warmer than the surrounding space, must begin to radiate forth its heat. We are to assume that the nebula does not receive heat from other bodies, adequate to compensate for that which it dissipates by radiation. There is thus a loss of heat and consequently the nebula must begin to contract. Its material must gradually draw together, and must do so under the operation of those fundamental laws which we have explained in the last chapter.

The contraction, or rather the condensation, of the material would of course generally be greatest at the central portion of the nebula. This is especially noticeable in the photograph of the great spiral already referred to. But in addition to this special condensation at the centre, the concentration takes place also, though in a lesser degree, at many other points throughout the whole extent of the glowing mass. Each centre of condensation which in this way becomes established tends continually to increase. In consequence of this law, as the great nebula contracted and as the great bulk of the material drew in towards the centre, there were isolated regions in the nebula which became subordinate centres of condensation. Perhaps in the primæval nebula, from which the solar system originated, there were half-a-dozen or more of these centres that were of conspicuous importance, while a much larger number of small points were also distinguished from the surrounding nebula. (Figures 40 and 41.) And still the contraction went on. The heat, or rather the energy with which the nebula had been originally charged, was still being dissipated by radiation. We give no estimate of the myriads of years that each stage of the mighty process must have occupied. The tendency of the transformation was, however, always in one direction. It did at last result in a great increase of the density of the substance of the nebula, both in the central regions as well as in the subordinate parts. In due time this increase in density had reached such a point that the materials in the condensing centres could be no longer described as retaining the gaseous form.

But though heat was incessantly being radiated from the great nebula, it did not necessarily follow that the nebula was itself losing temperature. This is a seeming paradox to which we have already had occasion to refer in Chapter 6. We need not now further refer to it than to remember that, in speaking of the loss of heat from the nebula, it would sometimes not be correct to describe the operation as that of cooling. Up to a certain stage in the condensation, the loss of heat leads rather to an augmentation of temperature than to its decline.

We are thus led to see how the laws of heat, after being in action on the primitive nebula for a period of illimitable ages, have at last effected a marvellous transformation. That nebula has condensed into a vast central mass with a number of associated subordinate portions. We may suppose that the original nebula in the course of time does practically disappear. It is absorbed by the attraction of those ponderous centres which have gradually developed throughout its extent.

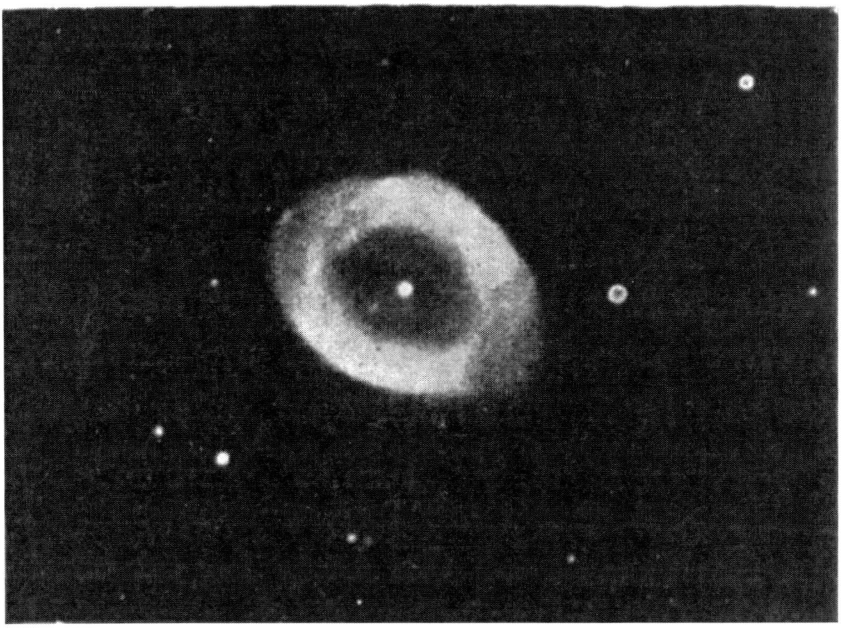

Figure 41. The Ring Nebula in Lyra (Lick Observatory).
(*From the Royal Astronomical Society Series.*)

The large central body, and perhaps some of the other bodies thus evolved, are at first of so high a temperature that a copious radiation of heat still goes forth from the system. As they discharge their stores of heat, the smaller bodies show the effects of loss of heat more rapidly than those which are larger. It is indeed obvious that a small body must cool more rapidly than a big one. It is sufficient to note that the cooling takes place from the surface, and that the bigger the body the larger the quantity of material that it contains for each unit of superficial area. If the radius of a sphere be doubled, its volume is increased eightfold, while its surface is only increased fourfold.

Let us now concentrate our attention on two of the bodies which, after immense ages, have been formed from the condensation of the primæval nebula. Let one of the two bodies be that central object, which preponderates so enormously that its mass is a thousandfold that of all the others taken together. Let the other be one of the smaller bodies. As it parts with its heat, the smaller body, which has originally condensed from the nebula, will assume some of the features of a mass of molten liquid. From the liquid condition, the body will pass with comparative rapidity into a solid state, at least on its outer parts. The exterior of this body will therefore become solid while the interior is still at an excessively high temperature. The outer material, which has assumed the solid form, is constituted of the elements with which we are acquainted, and is in the form of what the geologist would class as the igneous rocks, of which granite is the best known example. The shell of hard rocks outside encloses the material which is still heated and molten inside. Such a crust would certainly be an extremely bad conductor of heat. The internal heat is therefore greatly obstructed in its passage outwards to the surface. The internal heat may consequently be preserved in the interior of the body for an enormously protracted period, a period perhaps comparable with those immense ages which the evolution of the body from the primæval nebula has demanded. The smaller body may have thus attained a condition in which the temperature reigning on its surface is regulated chiefly by the external conditions of the space around, while the internal parts are still highly charged with the primitive heat from the original nebula.

The great central mass, which we may regard as thousands of times greater than that of the subordinate body, cools much more slowly. The cooling of this great mass is so enormously protracted in comparison with that of the smaller body that it is quite conceivable the central mass may continue to glow with intense fervour for immense ages after the smaller body has become covered with hard rock.

It will, I hope, be clear that the two bodies to which I am here alluding are not merely imaginary objects. The small body, which has so far cooled down that its surface has lost all indication of internal heat, is of course our earth. The great central mass which still glows with intense fervour is the sun. Such is in outline the origin of the sun and the earth as suggested by the nebular theory.

What we have said of the formation of the earth will equally apply to the evolution of other detached portions of the primitive nebula. There may be several of these, and they may vary greatly in size. The smaller they are the more rapidly in general will the superabundant heat be radiated away, and the sooner will the surface of that planet acquire the temperature which is determined by the surrounding conditions. There are, however, many modifying circumstances.

It is essential to notice that the primæval nebula must have had some initial moment of momentum, unless we are to assume the occurrence of that which is infinitely improbable. It would have been infinitely improbable for the system not to have had some moment of momentum originally. As the evolution proceeds, and as the energy is expended, while this original endowment of moment of momentum is preserved, we find, as explained in the last chapter, the system gradually settling down into proximity to a plane, and gradually acquiring a uniform direction of evolution. Hence we see that each of the subordinate masses which ultimately consolidate to form a planet have a motion of revolution around the central body. In like manner the central body itself rotates, and all these motions are performed in the same direction.

In addition to the revolutions of the planets around the sun, there are other motions which can be accounted for as consequences of the contraction of the nebula. We now refer to that central portion which is to form the sun, and consider, in the first instance, only one of the subordinate portions which is to form a planet. As these two bodies form part of the same nebulous mass they will to a certain extent rotate together as one piece. If anybody is rotating as a whole, every part of that body is also in actual rotation. We shall refer to this again later on; but for the present it is sufficient to observe that as the planet was originally continuous with the sun, it had a motion of rotation besides its motion of revolution, and it revolved round its own axis in a period equal to that of its revolution round the sun. In the beginning the rotation of the planet was therefore an exceedingly slow movement. But it became subsequently accelerated. For we have already explained that each planet is by itself subjected to the law of the conservation of moment of momentum. As each planet assumes a separate existence, it draws to itself its share of the moment of momentum, and that must be strictly preserved. But the planet, or rather the materials which are to form the future planet, are all the time shrinking; they are drawing more closely together. If, therefore, the area which each particle of the planet describes when multiplied by the mass of that particle and added to the similar products arising from all the other particles, is to remain constant, it becomes necessary that just as the orbits of these particles diminish in size, so must the speed at which they revolve increase. We thus find that there is a tendency in the planet to accelerate its rotation. And thus we see that a time will come when the planet, having assumed an independent existence, will be found rotating round its axis with a velocity which must be considered high in comparison with the angular velocity which the planet had while it still formed part of the original nebula.

As the planets have been evolved so as to describe their several orbits around the sun, so in like manner the smaller systems of satellites have been so evolved as to describe their orbits round the several planets that are their respective primaries. When a planet, or rather the materials which were drawing together to form a planet, had acquired a predominant attraction for the parts of the primæval nebula in their locality, a portion of the nebulous material became specially associated with the planet. As the planet with this nebulous material became separated from the central contracting sun, or became, as it were, left behind while the sun was drawing into itself the material which surrounded it the planet and its associated nebula underwent on a miniature scale an evolution similar to that which had already taken place in the formation of the sun and the planets as a whole. In this manner secondary systems seem sometimes to have had their origin.

We should, however, say that though what we have here indicated appears to explain fully the evolution of some of the systems, such, for instance, as that of Jupiter and his four moons, or Saturn and his eight or nine, the circumstances with regard to the earth and the moon are such as to require a very different explanation of the origin of our satellite. In the first place we may notice that the great mass of the moon, in comparison with the earth, is a wholly exceptional feature in the relations between the planets and their satellites in the other parts of the system. In no other instance does the mass of a satellite bear to the mass of the planet a ratio anything like so great as the ratio of our moon to the earth. The moon has a mass which is about one-eightieth of the mass of the earth, while even the largest of Jupiter's satellites has not one ten-thousandth part of the mass of the planet itself. The evolution of the earth and moon system has been brought about in a manner very different from that of the evolution of the other systems of satellites. We do not here enter into any discussion of the matter. We merely remind the reader that it is now known, mainly by the researches of Professor G. H. Darwin, that in all probability the moon was originally part of the earth, and that a partition having occurred while the materials of the earth and moon were still in a plastic state, a small portion broke away to form the moon, leaving behind the greater mass to form the earth. Then, under the influence of tides, which may agitate a mass of molten rock, as the moon was once, just as they may agitate an ocean, the moon was forced away, and was ultimately conducted to its present orbit.

It was at first tempting to imagine that a theory which accounted so satisfactorily for the evolution of the moon from the earth might also account in a similar manner for the evolution of the earth from the sun. Had this been the case, it is needless to say that the principles we now accept in the nebular theory would have needed large modification, if not actual abandonment. A close examination into the actual statistics brings forcibly before us the exceptional character of the earth-moon system. It can be demonstrated that the earth could not have been evolved from the sun in the same manner as there is every reason to believe that the moon has been evolved from the earth. The evolution of the satellites of Jupiter has proceeded along lines quite different from those of the evolution of the moon from the earth, so that we may, perhaps, find in the evolution of the satellites of Jupiter an illustration in miniature of the way in which the planets themselves have been evolved in relation to the sun.

We must not forget that the only spiral nebulæ which lie within the reach of our powers of observation, whether telescopic or photographic, appear to be objects of enormously greater cosmical magnificence than was that primæval nebula from which so insignificant an object as the solar system has sprung. The great spirals, so far as we can tell at present, appear to be thousands of times, or even millions of times, greater in area than the solar system. At this point, however, we must speak with special caution, having due regard to the paucity of our knowledge of a most important element. Astronomers must confess that no efforts which have yet been made to determine the dimensions of a nebula have been crowned with success.

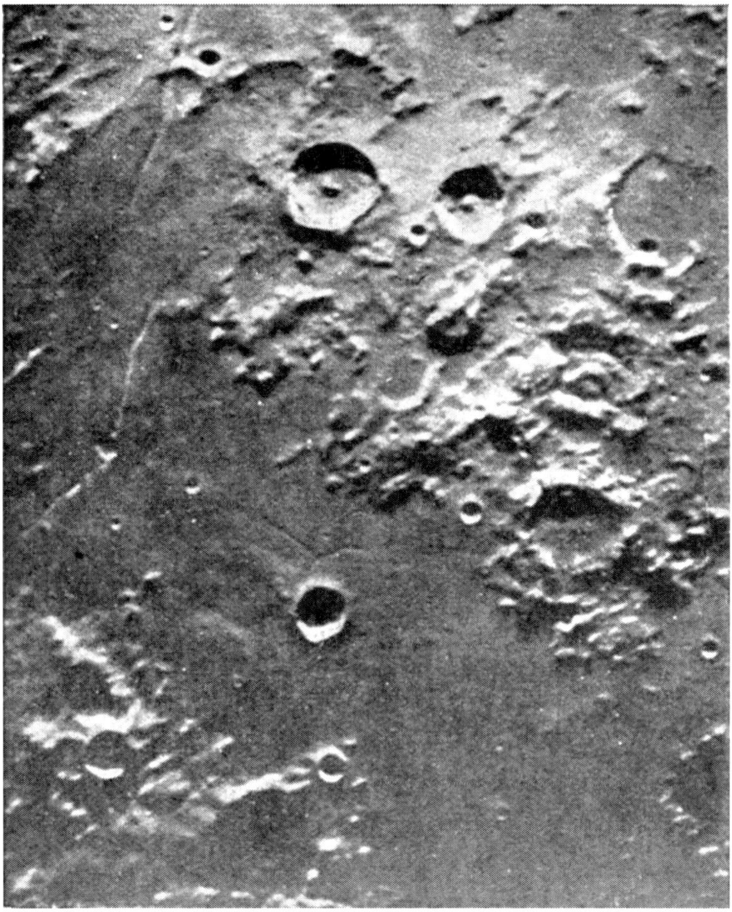

Figure 42. Lunar Craters: Hyginus and Albategnius.
(*Photographed by MM. Loewy and Puiseux.*)

We have not any precise idea as to what the distance of the great spiral might be. We generally take for granted that these nebulæ are at distances comparable with the distances of the stars. On this assumption we estimate that the spiral nebulæ must transcend enormously the dimensions of the primæval nebula from which the solar system has sprung. The spiral nebulæ that have so far come within our observation seem to be objects of an order of magnitude altogether higher than a solar system. They seem to be engaged on the majestic function of evolving systems of stars like the Milky Way, rather than on the inconsiderable task of producing a system which concerns only a single star and not a galaxy.

Figure 43. A remarkable Spiral (n.g.c. 628; in Pisces).
(Photographed by Dr. Isaac Roberts, F.R.S.)

The spiral form of structure is one in which Nature seems to delight. We find it in the organic world allied with objects of the greatest interest and beauty. The ammonite, a magnificent spiral shell sometimes exceeding three feet in diameter, belongs to a type which dominated the waters of the globe in secondary times, and which still survives in the nautilus. The same form is reproduced in minute creations totally different from ammonites in their zoological relations. Among the exquisite foraminifera which the microscopist knows so well may be found most delicate and beautiful spirals. Just as we see every range of spiral in the animal world, from an organism invisible to the naked eye, up to an ammonite a yard or more across, so it would seem that there are spiral nebulæ ranging from such vast objects as the great spiral in Canes Venatici down to such relatively minute spirals as those whose humble function it is to develop a solar system. It is no more than a reasonable supposition that the great spirals in the heavens are probably only the more majestic objects of an extremely numerous class. The smaller objects of this type—among which we might expect to find nebulæ like, in size and importance, to the primæval nebula of our system—are so small that they have not yet been recognised.

Figure 44. A clearly-cut Spiral (n.g.c. 4321; in Coma Berenices).
(Photographed by Dr. Isaac Roberts, F.R.S.)

It should at this stage be mentioned that several curious small planetary nebulæ have in these modern days been discovered by their peculiar spectra. If the nebulous character of these most interesting objects had not been accidentally disclosed by characteristic lines in their spectra, these undoubted nebulæ would each have been classified merely as stars. This fact will lead us to the surmise that there must be myriads of nebulæ in the heavens, too small to come within the range of our telescopes or of our most sensitive photographic plates. Suppose that a facsimile of the primæval nebula of our system, precisely corresponding with it in size and identical with it in every detail, were at the present moment located in space, but at a distance from our standpoint, as great as the distance of, let us say, the great spiral; it seems certain that this nebula, even though it contained the materials for a huge sun and a potential system of mighty planets, if not actually invisible to us here, would in all probability demand the best powers of our instruments to reveal it, and then it would be classed not as a nebula at all but as a star of perhaps the 12th or 15th, or even smaller magnitude.

It is to be remembered that the class of minute planetary nebulæ make themselves known solely by the fact that they exhibit the bright line indicative of gaseous spectra. If these objects (though still nebulæ) had not displayed gaseous spectra, it is certain they would have escaped detection, at least by the process which has actually proved so successful. The continuous band of light which they would then have presented could not be discriminated from the band of light from a star. It is therefore not improbable that among the star-like bodies which have been represented on our photographs, there may be some which are really minute spiral nebulæ. In general a star is a minute point of light which no augmentation of telescopic power and no magnification will show otherwise than as a point, granted only good optical conditions and good opportunity so far as the atmosphere is concerned. It has, however, been occasionally noted that certain so-called stars are not mere points of light; they do possess what is described as a disc. It is not at all impossible that the objects so referred to are spiral nebulæ. We may describe them as formed on a small scale in comparison with the great spiral or the nebula in Andromeda. But the smallness here referred to is only relative. They are in all probability quite as vast as the primæval spiral nebula from which the solar system has been evolved, though not so large as those curious ring-shaped nebulæ of which the most celebrated example lies in the constellation Lyra.

Such is an outline of what we believe to have been the history of our solar system. We have already given the evidence derived from the laws of heat. We have now to consider the evidence which has been derived from the constitution of the system itself. We shall see how strongly it supports the belief that the origin of sun and planets has been such as the nebular theory suggests.

Chapter 13

The Unity of Material in the Heavens and the Earth

In considering how the formation of our solar system was brought about, we naturally first enquire as to the material of which this superb scheme is constructed. What were the materials already to hand from which, in pursuance of the laws of Nature, the solar system was evolved?

See the robust and solid nature of this earth of ours, and the robust and solid nature of the moon and the planets. It might at first sight be concluded that the primitive materials of our earth had also been in the solid state. But such is not the case. The primitive material of the solar system was not solid, it was not even liquid. What we may describe as the mother-substance of the universe must have been of quite a different nature; we can give an illustration of the physical character of that substance.

The lover of Nature delights to look at the mountains and the trees, the lakes and the rivers. But he will not confine his regard merely to the objects on the earth's surface. He, no less than the artist and the poet, delights to gaze at that enchanting scenery which, day by day, is displayed in infinite beauty overhead; that scenery which is not wholly withheld even from observers whose lives may be passed amid the busy haunts of men, that scenery which is so often displayed on fine days at all seasons. We are alluding to those clouds which add the charm of infinite variety to the sky above us.

It is necessary for us now to think of matter when it possesses neither the density of a solid, nor the qualities of a liquid, but rather when it has that delicate texture which the clouds exhibit. The primæval material from which the solar system has been evolved is of a texture somewhat similar to that of the clouds. This primæval material is neither solid nor liquid; it is what we may describe as vapour.

But having pointed to the clouds in our own sky as illustrating, in a sense, the texture of this original mother-substance of the solar system, we can carry the analogy no further. Those dark and threatening masses which forbode the thunderstorm, or those beautiful fleecy clouds which enhance the loveliness of a summer's day, are, of course, merely the vapours of water. But the vapours in the mother-substance from which systems have been evolved were by no means the vapours of water. They were vapours of a very different character—vapours that suggest the abodes of Pluto rather than the gentle rain that blesses the earth. In the mother-substance of the solar system vapours of a great variety of substances were blended. For in the potent laboratory of Nature every substance, be it a metal or any other element, or any compound, no matter how refractory, will, under suitable circumstances, be dissolved into vapour.

Take, for instance, such a material as platinum. Could anything be less like a vapour than this silvery metal? We know that platinum is the densest of all the elements. We know that platinum, more effectually than other metals, resists liquefaction from the application of heat. No ordinary furnace can fuse platinum; yet in another way we can overcome the resistance of this metal. The electric arc, when suitably managed, yields a temperature higher than that of any furnace. Let the electric current spring from one pole of platinum to another, and a brilliant arc of light is produced by the glowing gas, which is characteristic of platinum. The light dispensed from that arc is different from the light that would be radiated if the poles were of any material other than platinum. Some of the platinum has not alone been melted, it has actually been turned into vapour by the overpowering heat to which it has been subjected. Thus the solidity of this substance, which resists so stubbornly the action of lower temperatures, can be overcome, and the very densest of all metals is dissolved into wisps of vapour.

We choose the case of platinum as an illustration because it is a substance exceptionally dense and exceptionally refractory. If platinum can be vaporised, there is not much difficulty in seeing that other elements must be capable of being vaporised also. In fact, given such heat as is found abundantly in natural sources, there is no known element, or combination of elements, which will not assume the form of gas or vapour or cloud.

At the temperature of the sun a drop of water would be forthwith resolved into its component gases of oxygen and hydrogen. In like manner a piece of chalk, if exposed to the sun, would be speedily transformed; it would first be heated red-hot and then white-hot; it is, indeed, white-hot chalk that gives us that limelight which we know so well. But the heat of the sun is far greater than the temperature of the incandescent lime. The lime would not only be heated white-hot by contact with solar heat, but still further stages would be reached. It would suffer decomposition. It would break up into three different elements: there would be the metal which we call calcium, there would be oxygen, and there would be carbon. Owing to the tremendous temperature of the sun the metal would not remain in the metallic form; it would not be even in a liquid form; it would become a gas. The elements which unite to form this chalk would be not only decomposed, but they would be vaporised. What is thus stated about the drop of water and the chalk may, so far as we know, be stated equally with regard to any other compounds. It matters not how close may be the chemical association in which the elements are joined: no matter how successfully those compounds may resist the decomposition under the conditions ordinarily prevailing on earth, they have to yield under the overwhelming trial to which the sun would subject them. Though there are many elements in the solar chemistry, there are no compounds. At the exalted temperature to which they are exposed in the sun the elements are indisposed for union with the other elements there met with, and which are at the same temperature. In these circumstances, they successfully resist all alliances.

Until the last few years no elements were known in our terrestrial experience which possessed at ordinary temperatures the same qualities of resolute isolation which all elements seem to display at extreme temperatures. The famous discovery of argon, and of other strange gases associated with argon in the atmosphere and elsewhere, has revealed, to the astonishment of chemists and to the great extension of knowledge, that we have with us here elements which resist all solicitations to enter into chemical union with other substances. It is doubtless in consequence of this absolute refusal to unite that, in spite of their abundance and their wide distribution, these elements have eluded detection for centuries. To the astronomer argon is both interesting and instructive. It shows us an element which possesses, at the ordinary temperatures of the surface of the earth, a property which is true of all elements when subjected to such temperatures as are found in the sun.

Think of the rocks which form the earth's crust and of the minerals which lie far below. Think of the soil which lies on its surface, of the forests which that soil supports, and the crops which it brings forth. Think of the waters of the ocean, and the ice of the Poles. Think of the objects of every kind on this globe. Think of the stone walls of a great building, of the iron used to give it strength, of the slates which cover it, and of the timber which forms its floors; think of the innumerable other materials which have gone towards its construction; think even of the elementary substances which go to form the bodies of animals, of the lime in their bones, and of the carbon which is so intimately associated with life itself. The nebular theory declares that those materials have not always been in the condition in which we now see them; that there was a time in which they were so hot that they were not in the solid state; they were not even in the fluid state, but were all in rolling volumes of glowing vapour which formed the great primæval fire-cloud.

We must understand the composite nature of the primitive fire-mist from which our solar system originated. Let me illustrate the matter thus: We shall suppose that a heterogeneous collection of substances is brought together, the items of which may be somewhat as follows: let there be many tons of iron and barrels of lime, some pieces of timber, and cargoes of flint; let there be lead and tin and zinc, and many other metals, from which copper and silver and several of the rarest metals must not be excluded; let there be innumerable loads of clay, which shall represent aluminium and silicon, and hogsheads of sea-water to supply oxygen, hydrogen, and sodium. There should be also, I need hardly add, many other elements; but there is no occasion to mention more; indeed, it would be impossible to give a list which would be complete.

Suppose that this diverse material is submitted to a heat as intense as the most perfect furnace can make it. Let the heat be indeed as great as that which we can get from the electric arc, or even greater still. Let us suppose this heat to be raised to such a point that, not only have the most refractory metals been transformed into vapour, but the elements which were closely in combination have also been rent asunder. This we know will happen when compound substances are raised to a very high temperature. We shall suppose that the heat has been sufficient to separate each particle of water into its constituent atoms of oxygen and hydrogen; we shall suppose that the heat has been sufficient to decompose even lime itself into its constituent parts, and exhibit them in the form of vapour. The heat is to be so great that even carbon itself, the most refractory of substances, has had to yield, so that after passing through a stage of dazzling incandescence it has melted and ultimately dissolved into vapour. Next let us suppose that these several vapours are blended, though we need not assume that the separate elements are diffused uniformly throughout all parts of the cloud. Let us suppose that these bodies, which contributed to form the nebula, have been employed in amounts, not to be measured in tons, or in hundreds of tons, but in a thousand millions of millions of millions of millions of tons. Let the mass of vapour thus arising be expanded freely through open space. Let it extend over a region which is to measure hundreds of thousands of millions of miles in length and breadth and depth. Then the doctrine of the earth's beginning, which we are striving to unfold in these lectures, declares that in a fire-mist such as is here outlined the solar system had its origin.

Various objections may occur to the thoughtful reader when asked to accept such statements. We must do our best to meet these objections. The evidence we submit must be of an indirect or circumstantial kind. Direct testimony on such a subject is from the nature of the case impossible. The actual fire-mist in which our system had its origin is a mist no longer. The material that forms the solid earth beneath our feet did once, we verily believe, float in the great primæval fire-mist. Of course we cannot show you that mist. Darwin could not show the original monkeys from which it would seem the human race has descended; none the less do most of us believe that our descent has really taken the line that Darwin's theory indicates.

In connection with this subject, as with most others, it is easy to ask questions which, I think we may say, no one can answer with any confidence. It may, for instance, be asked how this vast fire-mist came into existence. If it arose from heat, how did that heat happen to be present? Why was all the material in the state of vapour? What, in short, was the origin of that great primæval nebula? Here we must admit that we have proposed questions to which it is impossible for us to do more than suggest answers. As to what brought the mist into existence, as to whence the materials came, and as to whence the energy was derived which has been gradually expended ever since, we do not know anything, and, so far as I can see, we have no means of knowing. Conjectures on the subject are not wanting, of course, and in a later chapter we shall discuss what may be said on this matter.

I have shown you to some extent our reasons for believing that our solar system did originate in a fire-mist. And even if we are not able to explain how the mist itself arose, yet we do not admit that our argument as to the origin of our system is thereby invalidated. That such a fire-mist as the solar system required did once exist, must surely be regarded as not at all improbable so long as we can point to the analogous nebulæ or fire-mists which exist at the present moment, and which we see with our telescopes. Many of these are millions of times as great as the comparatively small fire-mist that would have evolved into our solar system.

A question has sometimes been asked as to the most important discovery in astronomy which has been made in the century that has just closed. If, by the most important discovery, we mean that which has most widely extended our knowledge of the Universe, I do not think there need be much hesitation in stating the answer. It seems to me beyond doubt that the most astonishing discovery of the last century in regard to the heavenly bodies is that which has revealed the elementary substances of which the orbs of heaven are composed. This discovery is the more interesting and instructive because it has taught us that the materials of the sun, of the stars, and of the nebulæ are essentially the elements of which our own earth is formed, and with which chemists had already become well acquainted.

We know, of course, that this earth, no matter how various may be the rocks and minerals which form its crust, and how infinite the variety of objects, organic and inorganic, which diversify its surface, is really formed from different combinations of about eighty different elements. There are gases like oxygen and hydrogen, there are other substances like carbon and sulphur, and there are metals like iron and copper. These elements are sometimes met with in their free or uncombined state, like oxygen in the atmosphere, or like gold in Klondike. More frequently they are found in combination, and in such combinations the characters of the constituent elements are sometimes completely transformed. A deadly gas and a curious metal, which burns as it floats on water, most certainly renounce their special characters when they unite to form the salt on our breakfast-table. Who would have guessed, if the chemist had not told him, that in every wheelbarrowful of ordinary earth there are pounds of silvery aluminium, and that marble is largely composed of an extremely rare metal, which but few people have ever seen?

Until the middle of the century just completed it seemed utterly impossible to form any notion as to the substances actually present in the sun. How could anyone possibly discern them by the resources of the older chemists? It might well have been doubted whether the elements of which the sun was made were the elements of which our earth was formed, and with which ordinary chemistry had made us familiar. Just as the animals and plants which met the gaze of the discoverers when they landed in the New World were essentially different from those in the Old World, so it might have been supposed, with good share of reason, that this great solar orb, ninety-three million miles distant, would be composed of elements totally different from those with which dwellers on the earth had been permitted to become acquainted.

This great discovery of the last century revealed to us the character of the elements which constitute the sun. It also added the astonishing information that they are essentially the same elements as those of which our earth itself and all which it contains are formed.

If anyone had asked in the early years of the century what those elements were which entered into the composition of the sun, the question would have been deemed a silly one; it would have been regarded as hopelessly beyond the possibility of solution, and it would have been as little likely to receive an answer as the questions people sometimes ask now as to the possible inhabitants on Mars.

But about the middle of the century a new era dawned; the wonderful method of spectroscopic analysis was discovered, and it became possible to examine the chemistry of the sun. The most important result was to show that the elements which enter into the composition of the sun are the same elements which enter into the composition of the earth. The student of the solar chemistry enjoys, however, one advantage over the terrestrial chemist, if it be an advantage to have his science simplified to the utmost extent. Chemistry would, however, lose its chief interest if all the elements remained as obstinately neutral as argon, and disdained alliance with all other elements. It would seem that those elements which most eagerly enter into combination here, and which resist with such vehemence our efforts to divorce them, must renounce all chemical union when exposed to the tremendous temperature of the sun.

Those elements which unite with the utmost eagerness at ordinary temperatures, seem to become indifferent to each other when subjected to the extremes of heat and cold. Potassium unites fiercely with oxygen in the most familiar of all chemical experiments. Potassium is indeed a strange metal, for it is of such small density that a piece cast on a basin of water will float like a chip of wood. It has such avidity for oxygen that it will decompose the water to wrench the molecules of oxygen from those of hydrogen. The union of the metal with the gas generates such heat that the strange substance bursts into flame. This is what takes place at the ordinary temperatures in the well-known experiment of the chemical lecture-table. But at extreme temperatures the greed of potassium for oxygen abates, if it does not vanish altogether. In those excessively low temperatures at which Professor Dewar experiments chemical affinities languish. He has reduced oxygen to a liquid, and he tells us that "a berg of silvery potassium might float forever untarnished on an ocean of liquid oxygen." At the excessively high temperature of the electric arc the oxygen and the potassium, whose union has been accomplished with such vehemence, cease to possess affinity, and they separate again.

The solar chemistry seems to know no combination. If a drop of water were transferred to the sun and subjected to the heat of the solar surface, it must immediately undergo decomposition. That which was a drop of water here would not remain a drop of water there; it would be at once resolved into its component elements of oxygen and hydrogen. The considerations just given greatly simplify the search for the particular bodies which are at present in the sun. We have only to test for the presence of each of eighty elements. We have not to take account of the thousands of chemical combinations of which these elements are susceptible under terrestrial conditions.

We are especially indebted to the late Professor Henry Rowland, of Baltimore, for a profound study of the solar spectrum. In his great work he enumerates thirty-six elements present in the sun, and the number may be increased now by at least two. Eight elements he classes as doubtful, fifteen are set down as absent from the solar spectrum, and several had not been tried. Iron stands foremost among all the solar elements, so far as the number of its lines are concerned. No fewer than 2,000 lines in the spectrum of the sun are attributed to this element. At the other end of the list lead is found. There is only one line apparently due to this metal. Carbon is represented by about 200 lines, and calcium by about 75. If, however, we test the significance of lines not by their number, but by their intensity, then iron no longer heads the list, its place being taken by calcium. Among the elements which Rowland sets down as not contributing any recognisable lines to the solar spectrum we may mention arsenic and sulphur, phosphorus, mercury, and gold.

Of the more prominent solar elements there are two or three of such special importance that we pause to give them a little consideration. Who does not remember the delight of the first occasion in childhood when he was permitted to peep into a bird's-nest and there see a group of eggs, often so exquisitely marked or so delicately tinted? How beautiful they seemed as they lay in their cosy receptacle concealed with so much cunning! Among other delightful recollections of early youth many will recall a ramble by the sea-shore. We may suppose the tide had retreated, and with other objects left by the sea on the gleaming sand a little cowrie shell is found. How enchanted we were with our prize! How we looked at the curious marks on its lips, and the inimitable beauty of its tints!

The shell of the hedge-sparrow and the shell cast up by the sea have another quality in common besides their beauty. They have both been fabricated from the same material. Lime is of course the substance from which the bird, by some subtle art of physiology, forms those exquisite walls by which the vital part of the egg is protected. The soft organism that once dwelt in the cowrie was endowed with some power by which it extracted from the waters of the ocean the lime with which it gradually built an inimitable shell. Is it an exaggeration to say that this particular element calcium, this element so excessively abundant and so rarely seen, seems to enjoy some peculiar distinction by association with exquisite grace and beauty? The white marble wrought to an unparalleled loveliness by the genius of a Phidias or a Canova is but a form of lime. So is the ivory on which the Japanese artist works with such delicacy and refinement. Whether as coral in a Pacific island, as a pearl in a necklace or as a stone in the Parthenon, lime seems often privileged to form the material basis of beauty in nature and beauty in art.

Though lime in its different forms, in the rocks of the earth or the waters of the ocean, is one of the most ordinary substances met with on our globe, yet calcium, the essential element which goes to the composition of lime, is, as we have already said, not by any means a familiar body, and not many of us, I imagine, can ever have seen it. Chemistry teaches that lime is the result of a union in definite proportions between oxygen gas and the very shy metal, calcium. This metal is never found in nature unless in such intimate chemical union with some other element like oxygen or chlorine, that its characteristic features are altogether obscured, and would indeed never be suspected from the mere appearance of the results of the union. To see the metal calcium you must visit a chemical laboratory where, by electrical decomposition or other ingenious process, this elusive element can be induced to part temporarily from its union with the oxygen or other body for which it has so eager an affinity, and to which it returns with such alacrity. Though calcium is certainly a metal, it is very unlike the more familiar metals such as gold or silver, copper or iron. A coin might conceivably be formed out of calcium, but it would have no stability like the coins of the well-known metals. Calcium has such an unconquerable desire to unite with oxygen that the unstable metal will speedily grasp from the surrounding air the vital element. Unless special precautions are taken to withhold from the calcium the air, or other source from whence it could obtain oxygen, the union will most certainly take place, and the calcium will resume the stable form of lime. Thus it happens that though this earth contains incalculable billions of tons of calcium in its various combinations, yet calcium itself is almost unknown except to the chemist.

It is plain that calcium plays a part of tremendous significance on this earth. I do not say that it is the most important of all the elements. It would indeed seem impossible to assign that distinction to any particular element. Many are, of course, of vital importance, though there are, no doubt, certain of the rarer elements with which this earth could perhaps dispense without being to any appreciable extent different from what it is at present. I do not know that we should be specially inconvenienced or feel any appreciable want unsatisfied, if, let us say, the element lanthanum were to be struck out of existence; and there are perhaps certain other rare bodies among the known eighty elements, about which the same remark might be made.

But without calcium there would neither be fertile soil for plants nor bones for animals, and consequently a world, inhabited in the same manner as our present globe, would be clearly impossible. There may be lowly organisms on this earth to which calcium is of no appreciable consequence, and it is of course conceivable that a world of living types could be constructed without the aid of that particular element which is to us so indispensable. But a world without calcium would be radically different from that world which we know, so that we are disposed to feel special interest in the important modern discovery that this same element, calcium, is abundantly distributed throughout the universe. The boldest and most striking features in the photograph of the solar spectrum are those due to calcium (Figure 42 and 44).

Figure 45. The H. and K. Lines in the Photographic Solar Spectrum (Higgs).

In the solar spectrum are two very broad, very dark, and very conspicuous lines, known as H and K. In every photograph of that portion of the solar spectrum which, lying beyond the extreme violet, is invisible to our eyes, though intensely active on the photographic plate, these lines stand forth so boldly as to arrest the attention more than any other features of the spectrum. It had been known that these lines were due to calcium, but there were certain difficulties connected with their interpretation. Some recent beautiful researches by Sir William and Lady Huggins have cleared away all doubt. It is now certain that the presence of these lines in the spectrum demonstrates that that remarkable element which is the essential feature of lime on this earth is also found in the sun. We have also to note that these same lines have been detected in the photographic spectra of many other bodies in widely different regions of space. Thus we establish the interesting result that this particular element which plays a part so remarkable on our earth is not restricted to our globe, but is diffused far and wide throughout the universe.

Perhaps the most astonishing discovery made in modern times about the sun is connected with the wonderful element, helium. So long ago as 1868 Sir Norman Lockyer discovered, during an eclipse, that the light of the sun contained evidence of the presence in that orb of some element which was then totally unknown to chemists. This new body was not unnaturally named the sun-element, or helium. But more than a quarter of a century had to elapse before any chemist could enjoy the opportunity of experimenting directly upon helium. No labour could prepare the smallest particle of this substance, no money could purchase it, for at that time no specimen of the element was known to exist nearer than the sun, ninety-three million miles distant. But in 1895 an astonishing discovery was made by Professor Ramsay. He was examining a rare piece of mineral from Norway. From this mineral, clevite, the Professor extracted a little gas which was to him and to all other chemists quite unknown. But on applying the spectroscope to examine the character of the light which this gas emitted when submitted to the electric current, it yielded, to their amazement, the characteristic light of helium. Thus was the sun-element at last shown to be a terrestrial body, though no doubt a rare one. The circumstances that I have mentioned make helium for ever famous among the constituents of the universe. It will never be forgotten that though from henceforth it may be regarded as a terrestrial body, yet it was first discovered, not in the earth beneath our feet, but in the far-distant sun.

In a previous picture we showed a photograph of a part of the sun's surface; this striking view displays those glowing clouds from which the sun dispenses its light and heat. These clouds form a comparatively thin stratum around the sun, the interior of which is very much darker. The layer of clouds is so thin that it may perhaps be likened to the delicate skin of a peach in comparison with the luscious interior. It is in these dazzling white clouds that we find the source of the sun's brightness. Were those clouds removed, though the sun's diameter would not be appreciably reduced, yet its unparalleled lustre would be at once lessened. We use the expression "clouds" in speaking of these objects, for clouds they certainly are, in the sense of being aggregates of innumerable myriads of minute beads of some substance; but those solar clouds are very unlike the clouds of our own sky, in so far as the material of which they are made is concerned. The solar clouds are not little beads of water; they are little beads of white-hot material so dazzlingly bright as to radiate forth the characteristic brilliance and splendour of the sun. The solar clouds drift to and fro; they are occasionally the sport of terrific hurricanes; they are sometimes driven away from limited areas, and in their absence we see merely the black interior of the solar globe, which we call a sun-spot. Now comes the important question as to the material present in these clouds which confers on the sun its ability to radiate forth such abundant light and heat.

The profound truth already stated, that the solar elements are the same as the terrestrial elements, greatly simplifies the search for that particular element which forms those solar clouds. As the sun is made of substances already known to us by terrestrial chemistry, and as there are no chemical compounds to embarrass us, the choice of the possible constituents of those solar clouds becomes narrowed to the list of elements experimented on in our laboratories.

We owe to Dr. G. Johnstone Stoney, F.R.S., the discovery of the particular element which forms those fire-clouds in the sun, and confers on the presiding body of the solar system the power of being so useful to the planets which owe it allegiance. Carbon is the element in question. I need hardly add that carbon is well known as one of the most commonplace and one of the most remarkable substances in Nature. A piece of coke differs from a piece of pure carbon only by the ash which the coke leaves behind when burned. Timber is principally composed of this same element, and when the timber is transformed into charcoal but little more than the carbon remains. Carbon is indeed everywhere present. It is, as we have mentioned, one of the elements which enter into the composition of a piece of chalk. Carbon is in the earth beneath our feet; it is in the air above us.

Carbon is one of the chief ingredients in our food, and it is by carbon that the heat of the body is sustained. Indeed, this remarkable element is intimately connected with life in every phase. Every organic substance contains carbon, and it courses with the blood in our veins. It assumes the widest variety of forms, renders the greatest diversity of services, and appears in the most widely different places. Carbon is indeed of a protean character, and there is a beautiful symbol of the unique position which it occupies in the scheme of Nature. Carbon is associated not alone with articles of daily utility and of plenteous abundance, but it is carbon which forms the most exquisite gems "of purest ray serene." The diamond is, of course, merely a specimen of carbon of absolute purity and in crystalline form. Great as is the importance of carbon on this earth, it is spread far more widely; it is not confined merely to the earth, for carbon abounds on other bodies in space. The most important functions of carbon in the universe are not those it renders on this earth. It was shown by Dr. Stoney that this same wonderful substance is indeed a solar element of vast utility. It is carbon which forms the glowing solar clouds to which our very life owes its origin.

In the incandescent lamp the brilliant light is produced by a glowing filament of carbon, and one reason why we employ this element in the electric lamp, instead of any other, may be easily stated. If we tried to make one of these lamps with an iron wire, we should find that when the electric current is turned on and begins to flow through the wire, the wire will, in accordance with a well-known law, become warm, then hot, red-hot, and white-hot; but even when white-hot the wire will not glow with the brightness that we expect from one of these lamps. Ere a sufficient temperature can be reached the iron will have yielded, it will have melted into drops of liquid, continuity will be broken, the circuit will be interrupted, and the lamp destroyed. We should not have been much more successful if instead of iron we had tried any other metal. Even a platinum wire, though it will admit of being raised to a much higher temperature than a wire of iron or a wire of steel, cannot remain in the solid condition at the temperature which would be necessary if the requisite incandescence is to be produced.

There is no known metal, and perhaps no substance whatever, which has so high a temperature of fusion as carbon. A filament of carbon, alone among the available elements, will remain continuous and unfused while transmitting a current intense enough to produce that dazzling brilliance which is expected from the incandescent lamp. This is the reason why this particular element carbon is an indispensable material for the electrician.

Modern research has now demonstrated that just as we employ carbon as the immediate agent for producing our beautiful artificial light, so the sun uses precisely the same element as the agent of its light and heat-giving power. In the extraordinary fervour which prevails in the interior of the sun all substances of every description must submit to be melted, nay, even to be driven into vapour. An iron poker, for instance, would vanish into iron vapour if submitted to this appalling solar furnace. Even carbon itself is unable to remain solid when subjected to the intense heat prevailing in the inner parts of the sun. At that heat carbon must assume the form of gas or vapour, just as iron or the other substances which yield more readily to the application of heat.

By the help of a simple experiment we may illustrate the significance of the carbon vapours in the solar economy. Let us take a Bunsen burner, in which the air and gas are freely mingled before they enter into combustion. If the air and the gas be properly proportioned, the combustion is so perfect that though a great deal of heat is produced there is but little light. The gas burned in this experiment ought to be the ordinary gas of our mains, which depends for its illuminating power on the circumstance that the hydrogen, of which the gas is chiefly composed, is largely charged with carbon. The illuminating power of the gas may indeed be measured by its available richness in carbon. As it enters the burner the carbon is itself in a gaseous form. This is not, of course, on account of a high temperature. The carbon of the coal-gas is in chemical union with hydrogen, and the result is in the form of invisible gases. It is these composite gases, blended with large volumes of ordinary hydrogen, which form the illuminating gas of our mains.

In the Bunsen burner the admission of a proper proportion of air, which becomes thoroughly mixed with the coal gas, produces perfect combustion. In the act of burning, the oxygen of the air unites immediately with the gas; it combines with the hydrogen to form watery vapour, and it combines with the carbon to form gases which are the well-understood products of combustion.

Suppose, now, we cut off the supply of air from the Bunsen burner, which can be done in a moment by placing the hand over the ring of holes at the bottom at which the air is admitted. Immediately a change takes place in the combustion. In place of the steady, hardly visible, but intensely hot flame which we had before, we have now a very much larger flame which makes a bright and flickering flare that lights up the room. If we re-admit the air at the bottom of the burner the light goes down instantly; the small, pale flame replaces it, and again the perfect combustion gives out intense heat at the expense of the light.

The remarkable change in the character of a gas-flame produced by admitting air to mix with the gas before combustion is, of course, easily explained. The chemical action takes place with much greater facility under these circumstances. The union of the carbon in the coal gas with the oxygen then takes place so thoroughly and instantaneously that the carbon never seems to have abandoned the gaseous form even for a moment in the course of the transformation. But in the case where air is not permitted to mingle with the gas, the supply of oxygen to unite with the incandescent gases can only be obtained from the exterior of the flame. The consequence is that the glowing gas charged with carbon vapour is chilled to some extent by contact with the cold air. It therefore seems as if the union of the hydrogen with the oxygen permitted the particles of carbon in the flame to resume their solid form for a moment. But in that solid form these particles, being at a high temperature, have a wonderful efficiency for radiation, and consequently brilliance is conferred upon the light. Most of the particles of carbon speedily unite with the surrounding oxygen, and re-enter the gaseous state in a different combination. Some of them, however, may escape this fate, in which case they assume the undesirable form of smoke. The Bunsen lamp can thus be made to give an illustration of the fact that when carbon vapours receive a chill, the immediate effect of the chill is to transform the carbon from the gaseous form to myriads of particles in the liquid, or more probably in the solid form. In the latter state the carbon possesses a power of radiation greatly in excess of that which it possessed in the gaseous state, even though the gas may have been at a much higher temperature than the white-hot solid particles.

We can now apply these principles to the explanation of the marvellous radiation of light and heat from the great orb of day. The buoyancy of the carbon vapours is one of their most remarkable characteristics; they tend to soar upwards through the solar atmosphere until they attain an elevation considerably over that of many of the other materials in the heated vapours surrounding the great luminary. We may illustrate what happens to these carbon vapours by considering the analogous case presented in the formation of ordinary clouds in our own skins. It is true, no doubt, that terrestrial clouds are composed of material very different from that which enters into the solar clouds. Terrestrial clouds of course arise in this way; the generous warmth of the sun evaporates water from the great oceans, and transforms it into vapour. This vapour ascends through our atmosphere, not at first as a visible cloud, but in the form of an invisible vapour. It is gradually diffused throughout the upper air, until at last particles of water, but recently withdrawn from the oceans, attain an altitude of a mile or more above the surface of the earth. A transformation then awaits this aqueous vapour. In the coldness of those elevated regions the water can no longer remain in the form of vapour. The laws of heat require that it shall revert to the liquid state. In obedience to this law the vapour collects into liquid beads, and it is these liquid beads, associated in countless myriads, which form the clouds we know so well. The same phenomenon of cloud-production is witnessed on a smaller scale in the formation of the visible puffs which issue from the funnel of a locomotive. We generally describe these rolling white volumes as steam; but this language is hardly correct. Steam, properly so called, is truly as invisible as the air itself; it is only after the steam has done its work and is discharged into the atmosphere, and there receives a chill, that it becomes suddenly transformed from the purely gaseous state into clustering masses of microscopic spheres of water, and thus becomes visible.

We can now understand the transformation of these buoyant carbon vapours which soar upwards in the sun. They attain an elevation at which the fearful intensity of the solar heat has been so far abated by the cold of outer space that the carbon gas is not permitted to remain any longer in the form of gas; it must return to the liquid or to the solid state. In the first stage on this return the carbon gas becomes transformed, just in the same way as watery vapour ascending from the earth becomes transformed into the fleecy cloud. Under the influence of its fall in temperature the carbon vapour collects into a clustering host of little beads of carbon. This is the origin of the glorious solar clouds. Each particle of carbon in that magnificent radiant surface has a temperature, and consequently a power of radiation, probably exceeding that with which the filament of carbon glows in the incandescent electric arc. When we consider that millions of millions of square miles on our luminary are covered with clouds, of which every particle is so intensely bright, we shall perhaps be able to form some idea of that inimitable splendour which even across the awful gulf of ninety-three million miles transmits the indescribable glory of daylight.

We are perhaps at present living rather too close to the period itself to be able to appreciate to its full extent the greatness of that characteristic discovery made in astronomy during the century just closed, to which the present chapter relates. In the early part of the last century it might have been said—indeed, by a certain very distinguished philosopher it actually was said—that a limit could be laid down bounding the possibilities of our knowledge of the heavenly bodies. It was admitted that we might study the movements of the different orbs in vastly greater detail than had been hitherto attempted, and that we might calculate the forces to which those orbs were submitted. With the help of mathematical analysis we might pursue the consequences of these forces to their remote ramifications; we might determine where the various orbs were situated at inimitably remote periods in the past. We might calculate the positions which they shall attain at epochs to be reached in the illimitably remote future; we might discover innumerable new stars and worlds; and we might map down and survey the distant parts of the universe. We might even sound the depths of space and determine the distances of the more remote celestial bodies, much more distant than any of those which have already yielded their secrets; we might measure the dimensions of those bodies and determine their weights; we might add scores or hundreds to the list of the known planets; we might multiply many times the number of known nebulæ and star-clusters; we might make measurements of many thousands of double stars; we might essay the sublime task of forming an inventory of the stars of the universe and compiling a catalogue in which the stars and their positions would be recorded in their millions; but, said the philosopher to whom I have referred, though you might accomplish all this, and much more in the same direction, yet there is a well-marked limit to your possible achievements; you can, he said, never expect to discover the actual chemical elements of which the heavenly bodies are composed. Nobody could dispute the reasonableness of this statement at the time he made it; indeed, it seemed to be a necessary deduction from our knowledge of the arts of chemistry, as those arts were understood before the middle of the last century.

In the prosecution of his researches by the older method, the chemist could no doubt discover the different elements of which the body was formed. That is to say, his art enabled him to accomplish this task, provided one very essential and fundamental condition could be complied with. However accomplished the chemist of fifty years ago might have been, he would assuredly have thought that he was being mocked if asked to determine the composition of a body which was 93,000,000 miles away from him. The very idea of forming an analysis under such conditions would have been scouted as preposterous. He would naturally ask that a specimen of the body should be delivered into his hands, a specimen which he could take into his laboratory, pulverise in his mortars, place in his test-tubes, treat with his re-agents, or examine with his blowpipe. Only by such methods was it then thought possible to obtain an analysis and discover the elements from which any given substance was formed.

For in the early part of this century the splendid method of spectrum analysis, that method which has revealed to us so many of the secrets of Nature, had not yet come into being. When that memorable event took place it was at once perceived that the spectroscope required no actual contact with the object to be tested, but only asked to receive some of the rays of light which that object dispersed when sufficiently heated. It was obvious that this new method must be capable of an enormously enlarged application. The flame producing the vapour might be at one end of the room, while the spectroscope testing the elements in that vapour might be at the other end. This new and beautiful optical instrument could analyse an object at a distance of a hundred feet. But if applicable at a distance of a hundred feet, why not at a hundred yards, or a hundred miles, or a hundred million miles? Why might the method not be used if the source of light were as far as the sun, or as far as a star, or even as far as the remotest nebula, whose faint gleam on the sky is all that the mightiest telescope can show.

Presently another great advance was recorded. As the study of this subject progressed, it was soon found that a spectrum visible to the human eye was not always indispensable for the success of the analysis. The photographic plate, which so frequently replaces the eye in other classes of observation, has also been used to replace the eye in the use of the spectroscope. A picture has thus been obtained showing the characteristic lines in the spectrum of a celestial object. That object may have been sunk in space to a distance so tremendous that even though the light travelled at a pace sufficient to complete seven circuits of our earth in each second of time, yet the rays from the object in question may have been travelling for centuries before they reached our instrument.

However the rays of light may have become weakened in the course of that journey, they still faithfully preserve the credentials of their origin. At last the light is decomposed in the spectroscope, and the several rays, which have been so closely commingled in their long voyage of myriads of miles, are now for the first time forced to pursue different tracks; they thus reach their different destinations on the photographic plate, and they there engrave their characteristic inscriptions. Nature in this operation imparts for our instruction a message which it is our business to interpret. It is true that these inscriptions are not always easily deciphered; many of them have not yet been understood. A portion of the solar spectrum showing many of the lines in the visible region is represented in the accompanying plate.

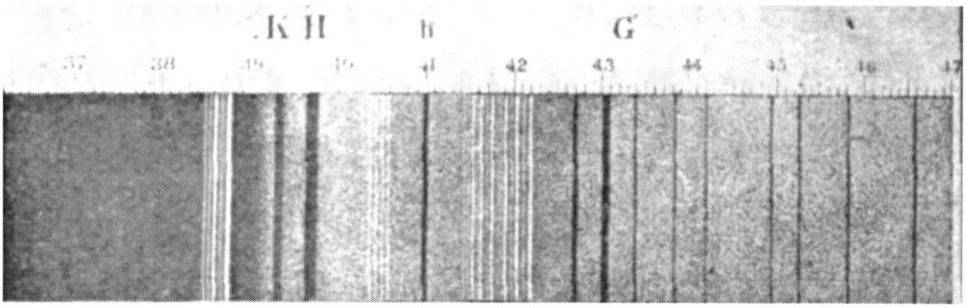

Figure 46. Spectrum of Comet showing Carbon Lines.
(Sir W. Huggins, K.C.B.)

Considering the insignificance of our earth when viewed in comparison with the millions of other orbs in the universe, considering also the stupendous distances by which the earth is separated from innumerable globes which are very much greater, it is certainly not a little astonishing to learn that the elements from which the various bodies in the universe have been composed are practically the same elements as those of which our earth is built. Is not this a weighty piece of evidence in favour of the theory that earth, sun, and planets are all portions of the same primæval nebula in which these elements were blended?

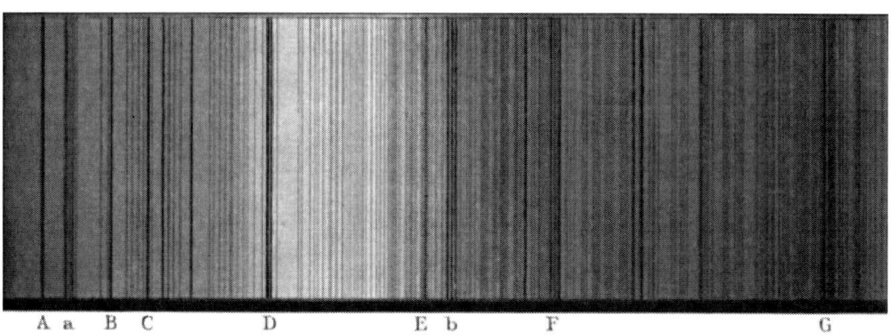

Figure 47. The Solar Spectrum.

We do not, of course, mean to affirm that the great primæval nebula was homogeneous throughout its vast extent. The waters of ocean are not strictly the same in all places; even the atmosphere is not absolutely uniform. Nature does not like homogeneity. The original nebula, we may well believe, was irregular in form, and denser in some places than in others. We do not suppose that if we could procure a sample of nebula in one place and another sample from the same nebula, but in a different place, say a hundred million miles distant, the two would show an identity of chemical composition; two samples of rock from different parts of the same quarry will not always be identical. But we may be assured that, in general, whatever elements are present in the nebula will be widely dispersed through its extent. If from different parts of the nebula two globes are formed by condensation, though we should not affirm, and though in fact we could not believe, that those globes would be of identical composition, yet we should reasonably expect that the elementary bodies which entered into their composition would be in substantial agreement. If one element, say iron, was abundant in one globe, we should expect that iron would not be absent from the other. Thus the elements represented in one body should be essentially those which were represented in the other.

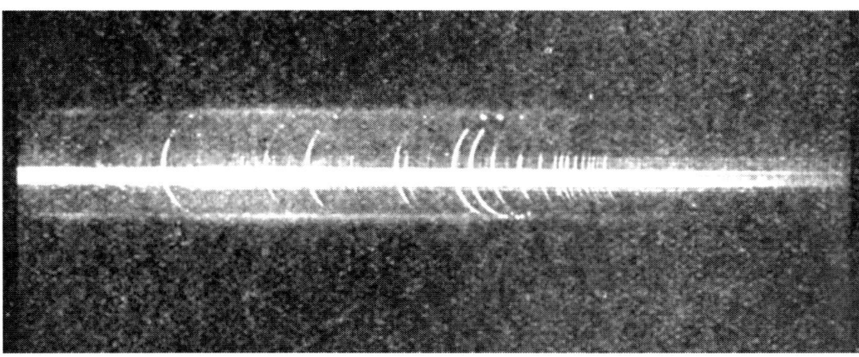

Figure 48. Spectrum of Sun during Eclipse. The Two Chief Lines are due to Calcium. (*Evershed.*)

It is obvious that if the sun and the earth—to confine our attention solely to those two bodies—had originated from the primæval nebula, they would bear with them, as a mark of their common origin, a resemblance in the elementary bodies of which they were composed. When Laplace framed his theory, he had not, he could not have had, the slightest notion as to the particular elements in the sun. For anything he could tell, those elements might be absolutely different from the elements in the earth. Yet, even without information on this critical point, the evidence for the nebular theory appeared to him so cogent that he gave it the sanction of his name.

It cannot be denied that if spectroscopic analysis had demonstrated that the elements in the sun were totally different from the elements in the earth a serious blow would have been dealt to the nebular theory. The collateral evidence, strong as it undoubtedly is, might hardly have withstood so damaging an admission. If, on the other hand, we find, as we actually have found, that the elements in the sun and the elements in the earth are practically identical, we obtain the most striking corroboration of the truth of the nebular theory. Had Kant and Laplace been aware of this most significant fact, they would probably have cited it as most important testimony. They would have pointed out that the iron so abundant in the earth beneath our feet is also abundant in the sun overhead. They would, I doubt not, if they had known it, have dwelt upon the circumstance that with that element, carbon, which enters into every organic body on this earth, our sun is also richly supplied, and they would have hardly failed to allude to the wide distribution in space of calcium, hydrogen, and many other well-known elements.

Laplace mainly based his belief in the nebular theory on some remarkable deductions from the theory of probabilities. To the consideration of these we proceed in the next three chapters. We may, however, remark at the outset that if the evidence derived from probabilities seemed satisfactory to Laplace one hundred years ago, this same line of evidence, strengthened as it has been by recent discoveries, is enormously more weighty, at the present day.

Chapter 14

THE FIRST CONCORD

In the present chapter, and in the two chapters which are to follow, I propose to give an outline of those arguments in favour of the nebular theory which are presented by certain remarkable coincidences observed in the movements of the bodies of our solar system. There are, indeed, certain features in the movements of the planets which would seem so inexplicable if the arrangement of the system had taken place by chance, that it is impossible not to seek for some physical explanation. We have already had occasion to refer in previous chapters to the movements of the bodies of our system. It will be our object at present to show that it is hardly conceivable that the movements could have acquired the peculiar characteristics they possess unless the solar system has itself had an origin such as that which the nebular theory assigns.

The argument on which we are to enter is, it must be confessed, somewhat subtle, but its cogency is irresistible. For this argument we are indebted to one of the great founders of the nebular theory. It was given by Kant himself in his famous essay.

We will commence with a preliminary point which relates to elementary mechanics. It may, however, help to clear up a difficult point in our argument if I now state some well-known principles in a manner specially adapted for our present purpose.

Let us think of two bodies, A and S, and, for the sake of clearness, we may suppose that each of these bodies is a perfect sphere. We might think of them as billiard balls, or balls of stone, or balls of iron. We shall, however, suppose them to be formed of material which is perfectly rigid. They may be of any size whatever, large or small, equal or unequal. One of them may be no greater than a grain of mustard-seed, and the other may be as large as the moon or the earth or the sun. Let us further suppose that there is no other body in the universe by which the mutual attraction of the two bodies we are considering can be interfered with. If these two bodies are abandoned to their mutual attraction, let us now see what the laws of mechanics assure us must necessarily happen.

Figure 49. A Spiral presented Edgewise (n.g.c. 4631; in Coma Berenices). *(Photographed by Dr. Isaac Roberts, F.R.S.)*

Let A and S be simply released from initial positions of absolute rest. In these circumstances, the two points will start off towards each other. The time that must elapse before the two bodies collide will depend upon circumstances. The greater the initial distance between the two balls, their sizes being the same, the longer must be the interval before they come together. The relation between the distance separating the bodies and the time that must elapse before they meet may be illustrated in this way. Suppose that two balls, both starting from rest at a certain distance, should take a year to come together by their mutual attraction, then we know that if the distance of the two balls had been four times as great eight years would have to elapse before the two balls collided. If the distances were nine times as great then twenty-seven years would elapse before the balls collided, and generally the squares of the times would increase as the cubes of the distances. In such statements we are supposing that the radii of the balls are inconsiderable in comparison with the distances apart from which they are started. The time occupied in the journey must also generally depend on the masses of the two bodies, or, to speak more precisely, on the sum of the masses of the two bodies. If the two balls each weighed five hundred tons, then they would take precisely the same time to rush together as would two balls of one ton and nine hundred and ninety-nine tons respectively, provided the distances between the centres of the two balls had been the same in each case. If the united masses of the two bodies amounted to four thousand tons, then they would meet in half the time that would have been required if their united masses were one thousand tons, it being understood that in each case they started with the same initial distance between the centres.

Instead of simply releasing the two bodies A and S so that neither of them shall have any impulse tending to make it swerve from the line directly joining them, let us now suppose that we give one of the bodies. A, a slight push sideways. The question will be somewhat simpler if we think of S as very massive, while A is relatively small. If, for instance, S be as heavy as a cannon-ball, while A is no heavier than a grain of shot, then we may consider that S remains practically at rest during the movement. The small pull which A is able to give will produce no more than an inappreciable effect on S. If the two bodies come together, A will practically do all the moving.

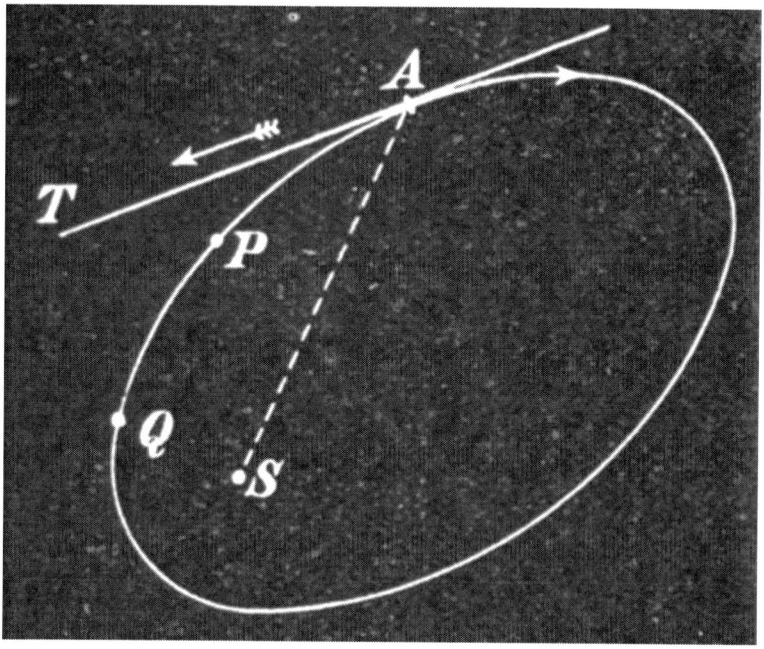

Figure 50. The Plane of a Planet's Orbit.

We represent the movement in the adjoining figure. If A is started off with an initial velocity in the direction A T, the attraction of S will, however, make itself felt, even though A cannot move directly towards S. The body will not be allowed to travel along A T; it will be forced to swerve by the attraction of S; it will move from P to Q, gradually getting nearer to S. To enter into the details of the movement would require rather more calculation than it would be convenient to give here. Even though S is much more massive than A, we may suppose that the path which A follows is so great that the diameter of the globe S is quite insignificant in comparison with the diameter of the orbit which the smaller body describes. We shall thus regard both A and S as particles, and Kepler's well-known law, to which we so often refer, tells us that A will revolve around S in that beautiful figure which the mathematician calls an ellipse. For our present purpose we are particularly to observe that the movement is restricted to a plane. The plane in which A moves depends entirely on the direction in which it was first started. The body will always continue to move in the same plane as that in which its motion originally commenced. This plane is determined by the point S and the straight line in which A was originally projected. It is essential for our argument to note that A will never swerve from its plane so long as there are not other forces in action beside those arising from the mutual attractions of A and S. The ordinary perturbations of one body by the action of others need not here concern us.

The case we have supposed will, of course, include that of the movement of a planet round the sun. The planet is small and represented by the body A, which revolves round the great body S, which stands for the sun. However the motion of the planet may actually have originated, it moves just as if it had received a certain initial impulse, in consequence of which it started into motion, and thus defined a certain plane, to which for all time its motion would be restricted.

So far we have spoken of only a single planet; let us now suppose that a second planet, B, is also to move in revolution about the same sun. This planet may be as great as A, or bigger, or smaller, but we shall still assume that both planets are inconsiderable in comparison with S. We may assume that B revolves at the same distance as A, or it may be nearer, or further. The orbit of B might also have been in the same plane as A, or—and here is the important point—it might have been in a plane inclined at any angle whatever to the orbit of A. The two planes might, indeed, have been perpendicular. No matter how varied may be the circumstances of the two planets, the sun would accept the control of each of them; each would be guided in its own orbit, whether that orbit be a circle, or whether it be an ellipse of any eccentricity whatever. So far as the attraction of the sun is concerned, each of these planets would remain forever in the same plane as that in which it originally started. Let us now suppose a third planet to be added. Here again we may assume every variety in the conditions of mass and distance. We may also assume that the plane which contains the orbit of this third planet is inclined at any angle whatever to the planes of the preceding planets. In the same way we may add a fourth planet, and a fifth; and in order to parallel the actual circumstance of our solar system, so far as its more important members are concerned, we may add a sixth, and a seventh, and an eighth. The planes of these orbits are subjected to a single condition only. Each one of them passes through the centre of the sun. If this requirement is fulfilled, the planes may be in other respects as different as possible.

In the actual solar system the circumstances are, however, very different from what we have represented in this imaginary solar system. It is the most obvious characteristic of the tracks of Jupiter and Venus, and the other planets belonging to the sun, that the planes in which they respectively move coincide very nearly with the plane in which the earth revolves. We must suppose all the orbits of our imaginary system to be flattened down, nearly into a plane, before we can transform the imaginary system of planets I have described into the semblance of an actual solar system.

If the orbits of the planets had been arranged in planes which were placed at random, we may presume they would have been inclined at very varied angles. As they are not so disposed, we may conclude that the planes have not been put down at random; we must conclude that there has been some cause in action which, if we may so describe it, has superintended the planes of these orbits and ordained that they should be placed in a very particular manner.

Two planets' orbits might conceivably coincide or be perpendicular, or they might contain any intermediate angle. The plane of the second planet might be inclined to the first at an angle containing any number of degrees. To make some numerical estimate of the matter, we proceed as follows: If we divide the right angle into ten parts of nine degrees each, then the inclination of the two planes might, for example, lie between O° and 9°, or between 18° and 27°, or between 45° and 54°, or between 81° and 90°, or in any one of the ten divisions. Let us think of the orbit of Jupiter. Then the inclination of the plane in which it moves to the plane in which the earth moves must fall into one of the ten divisions. As a matter of fact, it does fall into the angle between 0° and 9°.

But now let us consider a second planet, for example Venus. If the orbit of Venus were to be placed at random, its inclination might with equal probability lie in any one of the ten divisions, each of nine degrees, into which we have divided the right angle. It would be just as likely to lie between forty-five and fifty-four, or between seventy-two and eighty-one, as in any other division. But we find another curious coincidence. It was already remarkable that the plane of Jupiter's orbit should have been included in the first angle of nine degrees from the orbit of the earth. It is therefore specially noteworthy to find that the planet Venus follows the same law, though each one of the ten angular divisions was equally available.

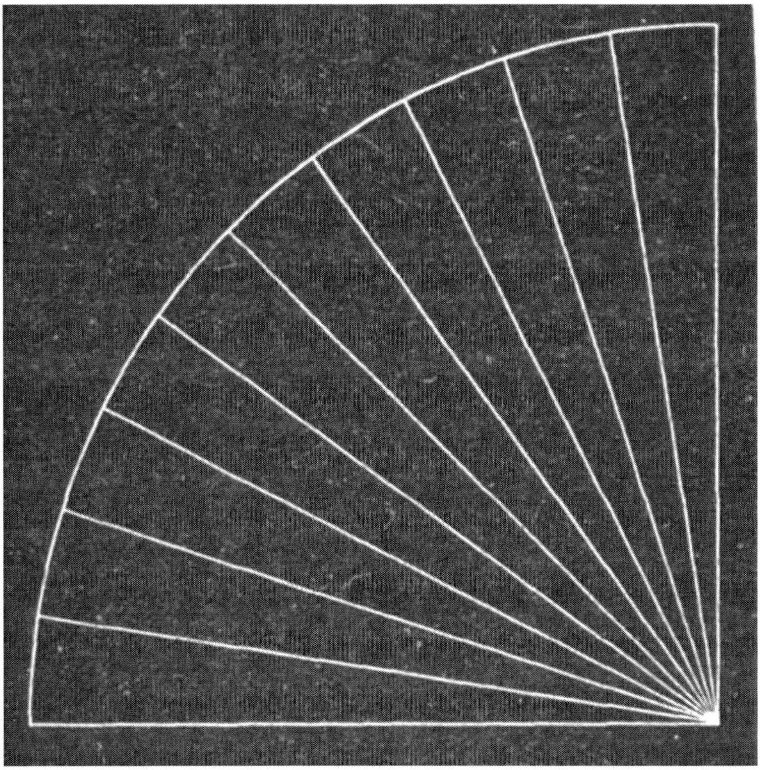

Figure 51. A Right Angle Divided into Ten Parts.

The coincidences we have mentioned, remarkable as they are, represent only the first of the series. What has been said with respect to the positions of the orbits of Jupiter and Venus may be repeated with regard to the orbits of Mercury and Mars, Saturn, Uranus, and Neptune. If the tracks of these planets had been placed merely at random, their inclinations would have been equally likely to fall into any of the ten divisions. As a matter of fact, they all agree in choosing that one particular division which is adjacent to the track of the earth. If the orbits of the planets had indeed been arranged fortuitously, it is almost inconceivable that such coincidences could have occurred. Let me illustrate the matter by the following little parable.

There were seven classes in a school, and there were ten boys in each class. There was one boy named Smith in the first class, but only one. There was also one Smith, but only one, in each of the other classes. The others were named Brown, Jones, Robinson, etc. An old boy, named Captain Smith, who had gone out to Australia many years before, came back to visit his old school. He had succeeded well in the world, and he wanted to do something generous for the boys at the place of which he had such kindly recollections. He determined to give a plum-cake to one boy in each class; and the fortunate boy was to be chosen by lot. The ten boys in each class were to draw, and each successful boy was to be sent in to Captain Smith to receive his cake.

The Captain sat at a table, and the seven winners were shown in to receive their prizes. "What is your name?" he said to the boy in the first class, as he shook hands with him. "Smith," replied the boy. "Dear me," said the Captain, "how odd that our names should be the same. Never mind, it's a good name. Here's your cake. Good-bye, Smith." Then up came the boy from the second class. "What is your name?" said the Captain. "Smith, sir," was the reply. "Dear me," said the visitor. "This is very singular. It is indeed a very curious coincidence that two Smiths should have succeeded. Were you really chosen by drawing lots?" "Yes, sir," said the boy. "Then are all the boys in your class named Smith?" "No, sir; I'm the only one of that name in the ten." "Well," said the Captain, "it really is most curious. I never heard anything so extraordinary as that two namesakes of my own should happen to be the winners. Now then for the boy from class three." A cheerful youth advanced with a smile. "Well, at all events," said the good-natured old boy, "your name is not Smith?" "Oh, but it is," said the youth. The gallant Captain jumped up, and declared that there must have been some tremendous imposition. Either the whole school consisted of Smiths, or they called themselves Smiths, or they had picked out the Smiths. The four remaining boys, still expecting their cakes, here burst out laughing. "What are your names?" shouted the donor. "Smith!" "Smith!!" "Smith!!!" "Smith!!!!" were the astounding replies. The good man could stand this no longer. He sent for the schoolmaster, and said, "I particularly requested that you would choose a boy drawn by lot from each of your seven classes, but you have not done so. You have merely picked out my namesakes and sent them up for the cakes." But the master replied, "No, I assure you, they have been honestly chosen by lot. Nine black beans and one white bean were placed in a bag; each class of ten then drew in succession, and in each class it happened that the boy named Smith drew the white bean."

"But," said the visitor, "this is not credible. Only once in ten million times would all the seven Smiths have drawn the white beans if left solely to chance. And do you mean to tell me that what can happen only once out of ten million times did actually happen on this occasion—the only occasion in my life on which I have attempted such a thing? I don't believe the drawing was made fairly by lot. There must have been some interference with the operation of chance. I insist on having the lots drawn again under my own inspection." "Yes, yes," shouted all the other boys. But all the successful Smiths roared out, "No." They did not feel at all desirous of another trial. They knew enough of the theory of probabilities to be aware that they might wait till another ten million fortunate old boys came back to the school before they would have such luck again. The situation came to a deadlock. The Captain protested that some fraud had been perpetrated, and in spite of their assurances he would not believe them. The seven Smiths declared they had won their cakes honestly, and that they would not surrender them. The Captain was getting furious, the boys were on the point of rebellion, when the schoolmaster's wife, alarmed by the tumult, came on the scene. She asked what was the cause of the disturbance. It was explained to her, and then Captain Smith added that by mathematical probabilities it was almost inconceivable that the only seven Smiths in the school should have been chosen. The gracious lady replied that she knew nothing, and cared as little, about the theory of probabilities, but she did care greatly that the school should not be thrown into tumult. “There is only one solution of this difficulty,” she added. “It is that you forthwith provide cakes, not only for the seven Smiths, but for every one of the boys in the school.” This resolute pronouncement was received with shouts of approval. The Captain, with a somewhat rueful countenance, acknowledged that it only remained for him to comply. He returned, shortly afterwards, to his gold-diggings in Australia, there to meditate during his leisure on this remarkable illustration of the theory of probabilities.

This parable illustrates the improbability of such arrangements as we find in the planets having originated by chance. The chances against their having thus occurred are 10,000,000 to 1. Hence we find it reasonable to come to the conclusion that the arrangement, by which the planets move round the sun in planes which are nearly coincident, cannot have originated by chance. There must have been some cause which produced this special disposition. We have, therefore, to search for some common cause which must have operated on all the planets. As the planets are at present absolutely separated from each other, it is impossible for us to conceive a common cause acting upon them in their present condition. The cause must have operated at some primæval time, before the planets assumed the separate individual existence that they now have.

We have spoken so far of the great planets only, and we have seen how the probability stands. We should also remark that there are also nearly 500 small planets, or asteroids, as they are more generally called. Among them are, no doubt, a few whose orbits have inclinations to the ecliptic larger than those of the great planets. The great majority of the asteroids revolve, however, very close to that remarkable plane with which the orbits of the great planets so nearly coincide. Every one of these asteroids increases the improbability that the planes of the orbits could have been arranged as we find them, without some special disposing cause. It is not possible or necessary to write down the exact figures. The probability is absolutely overwhelming against such an arrangement being found if the orbits of the planets had been decided by chance, and chance alone.

We may feel confident that there must have been some particular circumstances accompanying the formation of the solar system which rendered it absolutely necessary for the orbits of the planets to possess this particular characteristic. We have pointed out in Chapter 12. that the nebular theory offers such an explanation, and we do not know of any other natural explanation which would be worthy of serious attention. Indeed, we may say that no other such explanation has ever been offered.

Chapter 15

THE SECOND CONCORD

We have seen in the last chapter that there is a very remarkable concordance in the positions of the planes of the orbits of the planets, and we have shown that this concordance finds a natural historical explanation in the nebular origin of our system. We have now to consider another striking concord in the movements of the planets in their several orbits, and this also furnishes us with important evidence as to the truth of the nebular theory. The argument on which we are now to enter is one which specially appealed to Laplace, and was put forward by him as the main foundation of the nebular theory.

In the adjoining we have a diagram of a portion of the solar system. We shall regard the movements as somewhat simplified. The sun is supposed to be at the centre, turning round once every twenty-five days, on an axis which is supposed to be perpendicular to the plane of the paper. We may also for our present purpose assume that the orbits of the earth and the other planets lie in this same plane.

In the first place we observe that the earth might have gone round its track in either direction so far as the welfare of mankind is concerned. The succession of day and night, and the due changes of the seasons, could have been equally well secured whichever be the direction in which the earth revolves. We do, however, most certainly find that the direction in which the earth revolves round the sun is the same as the direction in which the sun rotates on its axis. This is the first coincidence.

We may now consider other planets. Look, for instance, at the orbit of Jupiter. It seems obvious that Jupiter might have been made to revolve round the sun either one way or the other; indeed, it will be remembered that though Kepler's laws indicate so particularly the shape of the track in which the planet revolves, and prescribe so beautifully the way in which the planet must moderate or accelerate its velocity at the different parts of its track, yet they are quite silent as to the direction in which the planets shall revolve in that track. If we could imagine a planet to be stopped to have its velocity reversed, and then to be started in a precisely opposite direction, it would still continue to describe precisely the same path; it would still obey Kepler's laws with unfailing accuracy, so far as our present argument is concerned, and the velocity which it would have at each point of the track would be quite the same whether the planet were going one way or whether it was going the other. It is therefore equally possible for Jupiter to pursue his actual track by going round the sun in the same direction as the earth, or by going in the opposite direction. But we actually find that Jupiter does take the same direction as the earth, and this, as we have already seen, is the direction in which the sun rotates. Here we have the second coincidence.

We now take another planet; for example, Mars. Again we affirm that Mars could have moved in either direction, but, as a matter of fact, it pursues the same direction as Jupiter and the earth. In the orbital movement of Saturn we have the fourth coincidence of the same kind, and we have a fifth in the case of Mercury, and a sixth in Venus, a seventh in Uranus, and an eighth in Neptune. The seven great planets and the earth all revolve around the sun, not only in orbits which are very nearly in the same plane, but they also revolve in the same direction.

The coincidences we have pointed out with regard to the movements of the great planets of our system may be also observed with regard to the numerous bodies of asteroids. On the first night of the century just closed, the 1st of January, 1801, the first of the asteroids, now known as Ceres, was discovered. This was a small planet, not a thousandth part of the bulk of one of the older planets, and visible, of course, only in the telescope.

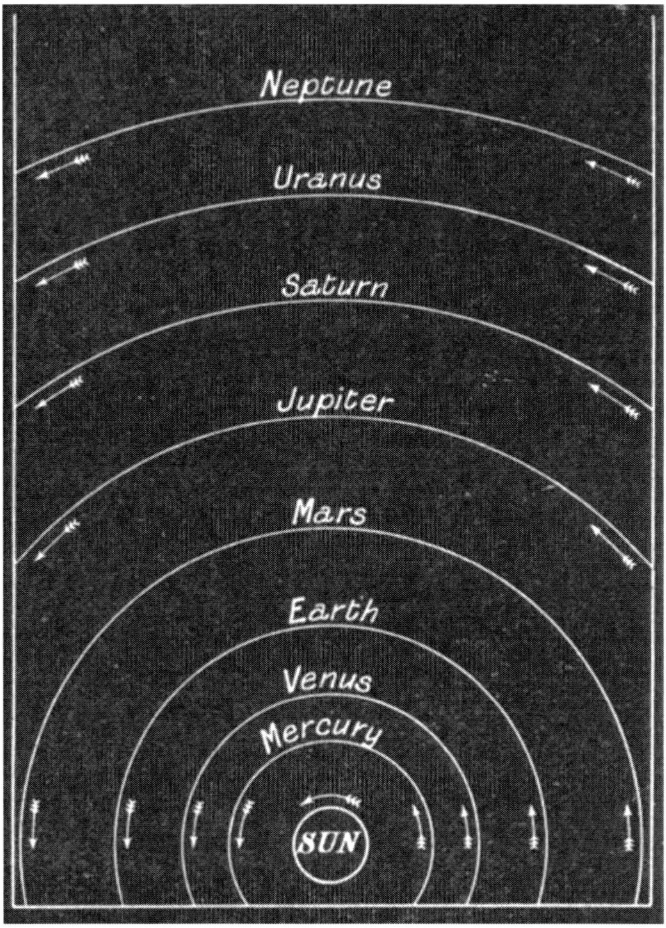

Figure 52. Illustration of the Second Concord.

Like the older planets, it was found to obey Kepler's laws; but this we might have foreseen, because Kepler's laws depend upon the attraction of gravitation, and must apply to any planet, whatever its size. When, therefore, the new planet was found, and its track was known, it was of much interest to see whether the planet in moving round that track observed the same direction in which all the older planets had agreed to travel, or whether it moved in the opposite direction. In the orbit of Ceres we have a repetition of the coincidence which has been noticed in each of the other planets. The new planets, like all the rest, move round the sun in the same direction as the sun rotates on its axis. The discovery of this first asteroid was quickly followed by other similar discoveries; each of the new planets described, of course, an ellipse, and the directions which these planets followed in their movements round the sun were in absolute harmony with those of the older planets.

But, besides the great planets and the asteroids properly so called, there is yet another planet, Eros. Its testimony is of special value, inasmuch as it seems to stand apart from all other bodies in the solar system, and with, of course, the exception of the moon, it is the earth's nearest neighbour. But whatever may be the exceptional features of Eros, however it may differ from the great planets and the asteroids already known, yet Eros makes no exception to the law which we have found to be obeyed by all the other planets. It also revolves round the sun in the same direction as all the planets revolve, in the same direction as the rotation of the sun.

We may pause at this moment to make a calculation as to the improbability that the sun, the earth, the seven great planets, and Ceres, numbering altogether ten, should move round in the same direction if their movements had been left to chance. This will show what we can reasonably infer from this concord in their movements. The theory of probabilities will again enlighten a difficult subject.

There are only two possible directions for the motion of a planet in its orbit. It must move like the hands of a watch, or it must move in the opposite direction. The planet must move one way or the other, just as a penny must always fall head or tail.

We may illustrate this remarkable coincidence in the following manner: Suppose we take ten coins in the hand, and toss them all up together and let them fall on the table; in the vast majority of cases in which the experiment may be tried, there would be some heads and some tails; they would not all be heads. But it is, of course, not impossible that the coins should all turn up heads. We should, however, deem it a very remarkable circumstance if it happened: yet it would certainly not be more remarkable than that the ten celestial movements should all take place in the same direction, unless, indeed, it should turn out that there is some sound physical cause which imposes on the planets of the solar system an obligation, restricting their movements round the sun to the same direction as that in which the sun itself rotates.

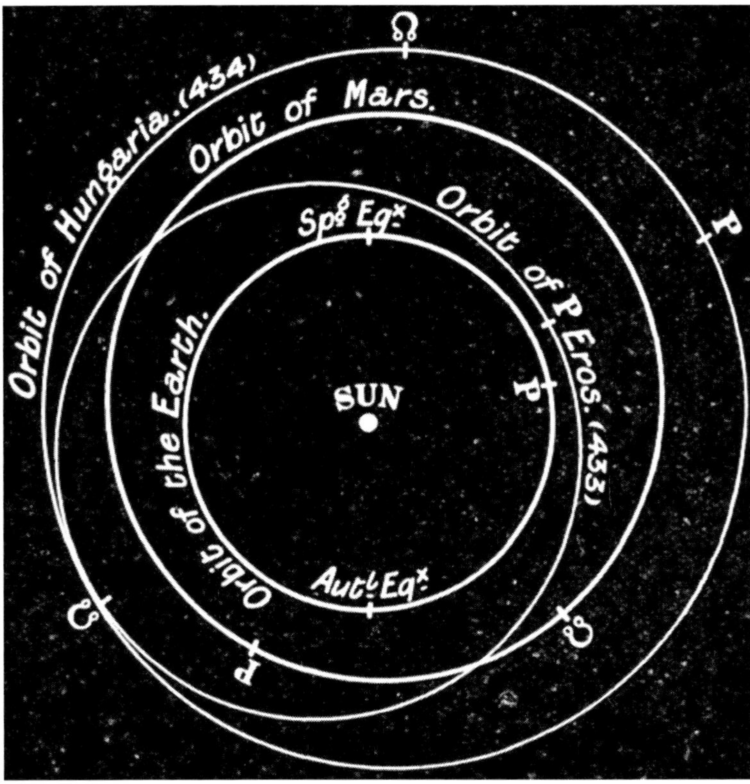

Figure 53. Orbits of Earth, Eros and Mars.

It will be useful to study the matter numerically; and the rules of probabilities will enable us to do so, as we may see by the following illustration: We deem the captain of a cricket team fortunate when he wins the toss for innings. We should deem him lucky indeed if he won it three times in successive matches. If he won it five times running, his luck would be phenomenal; while, if it was stated that he won it ten times consecutively, we should consider the statement well-nigh incredible. For it is easy to calculate that the chances against such an occurrence are one thousand and twenty-four to one. In like manner we may say, that for nine planets and the sun all to go round in the same direction would be indeed surprising if the arrangement of the planets had been determined by chance; there are more than a thousand chances to one against such an occurrence.

But Ceres was only the earliest of many other similar discoveries. And as each asteroid was successively brought to light, it became most interesting to test whether it followed the rest of the planets in that wonderful unanimity in the direction of their movements of revolution, or whether it made a new departure by going in the opposite direction. No such exception has ever yet been observed. Let us take, then, ten more planets, in addition to those we have already considered, so that we have now nineteen planets all revolving in the same direction as the sun rotates. It is easy to compute the improbability that these twenty movements should all be in the same direction, if, indeed, it were by chance that their directions had been determined. It is the same problem as the following: What is the chance that twenty coins, taken together in the hand and tossed into the air at once, shall all alight with their heads uppermost? We have seen that the chances against this occurrence, if there were ten coins, is about a thousand to one. It can easily be shown that if there were twenty coins the chances against the occurrence would be a million to one. We thus see that, even with no more than nineteen planets and the sun, there is a million to one against a unanimity in the directions of the movements, if the determination of the motions was made by chance. We may, however, express the result in a different manner, which is more to the purpose of our argument. There are a million chances to one in favour of the supposition that the disposition of the movements of the planets has not been the result of chance; or we may say that there are a million chances to one in favour of the supposition that some physical agent has caused the unanimity.

We can add almost any desired amount of numerical strength to the argument. The discoveries of minor planets went on with ever-increasing success through the whole of the last century. When ten more had been found, and when each one was shown to obey the same invisible guide as to the direction in which it should pursue its elliptic orbit, the chances in favour of some physical cause for the unanimity became multiplied by yet another thousand. The probability then stood at a thousand millions to one. As the years rolled by, asteroids were found in ever-increasing abundance. Sometimes a single astronomer discovered two, and sometimes even more than two, on a single night. In the course of a lifetime a diligent astronomer has placed fifty discoveries of asteroids, or even more than fifty, on his record. By combined efforts the tale of the asteroids has now approached five hundred, and out of that huge number of independent planetary bodies there is not one single dissentient in the direction of its motion. Without any exception whatever, they all perform their revolutions in the same direction as the sun rotates at the centre. When this great host is considered, the numerical strength of the argument would require about 150 figures for expression. Each new asteroid simply doubled the strength of the argument as it stood before.

Professor J. J. Thomson recently discovered that there are corpuscles of matter very much smaller than atoms. Let us think of one of these corpuscles, of which many millions would be required to make the smallest grain of sand which would just be visible under a microscope. Think, on the other hand, of a sphere extending through space to so vast a distance that every star in the Milky Way will be contained within its compass. Then the number of those corpuscles which would be required to fill that sphere is still far too small to represent the hugeness of the improbability that all the five hundred planetary bodies should revolve in the same direction, if chance, and chance alone, had guided the direction which each planet was to pursue in moving round its orbit.

The mere statement of these facts is sufficient to show that some physical agent must have caused this marvellous concord in the movements of the solar system. How the argument would have stood if there had been even a single dissentient it is not necessary to consider, for there is no dissentient No reasonable person will deny that these facts impose an obligation to search for the physical explanation of this feature in the planetary movements.

As in the last chapter, where we were dealing with the positions of the planes of the orbits, there can here be no hesitation as to the true cause of this most striking characteristic of the planetary movements. The nebular theory is at once ready with an explanation, as has been already indicated in Chapter 11. The primæval nebula, endowed in the beginning with a certain amount of moment of momentum, has been gradually contracting. It has been gradually expending its energy, as we have already had occasion to explain; but the moment of momentum has remained undiminished. And from this it can be shown that the dynamical principles guiding the evolution of the nebula must ultimately refuse permission for any planet to revolve in opposition to the general movement. This point is a very interesting one, and as it is of very great importance in connection with our system, I must give it some further illustration and explanation.

The two figures that are shown in Figure 54 represent two imaginary systems. We have a sun in each, and we have two planets in each. The sun is marked with the letter S, and the two planets are designated by A and B. For simplicity I have represented the orbits as circles, and for the same reason I have left out the rest of the planets; we shall also suppose the orbits of the two planets that are involved to lie exactly in the same plane. In the two systems that I have here supposed, the two suns are to be of the same weight, the planet A in one system is of equal mass to the planet A in the other; and the planets B in the two systems are also equal. It is also assumed that the orbit of A in one diagram shall be the same as the orbit of A in the other, and that the orbit of B in one shall be precisely the same as the orbit of B in the other. The sun rotates in precisely the same manner in both, and takes the same time for each rotation. A, in one system, goes round in the same time that A does in the other; and B, in one system, goes round in the same time that B does in the other. There is, therefore, a perfect resemblance between the two systems I have here supposed in every point but one. I have indicated, as usual, the movements of the bodies by arrows, and, while in one of the systems the sun and A and B all go round in the same direction, in the other system the sun and A go round, no doubt, in the same direction, but the direction of B is opposite. We are not, in this illustration, considering the rotations of the planets on their axes. That will be dealt with in the next chapter.

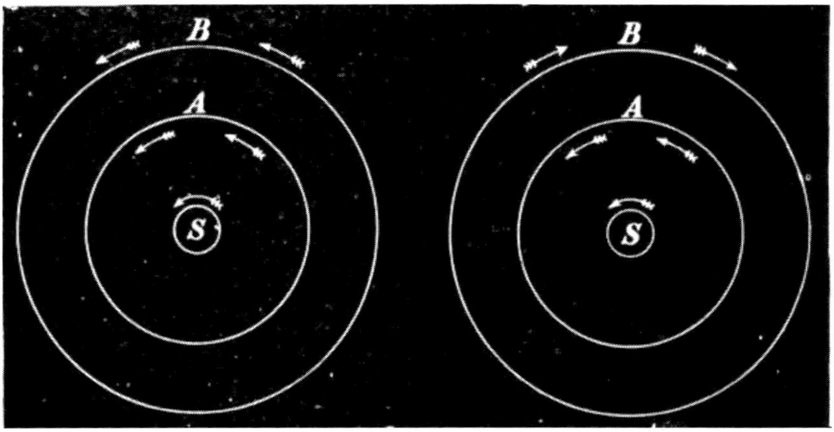

Figure 54. I. A Natural System on the Left. II. An Unnatural System on the Right.

There can be no doubt that either of these two systems would be possible for thousands of revolutions. There is nothing whatever to prevent A and B from being started in the same direction round the sun as in the first figure, or with A in one direction and B in the opposite direction, as in the second figure. It is equally conceivable that, while A and B revolve in the same direction, both should be opposite to that of the sun. But one system is permanent, and the other is not.

For, as a matter of fact, we do not find in Nature such an arrangement as that in the second figure, or as that in which both the planets revolve in opposite directions to the sun's rotation; what we do find is, that the planets go round in the same direction as the sun. And the explanation is undoubtedly connected with the important principle already illustrated, namely, that natural systems are in a condition in which the total quantity of energy undergoes continuous reduction in comparison with the moment of momentum.

In the arrangements made in the two figures, it will be recollected that the masses of the three bodies were respectively the same, and also their distances apart, and their velocities. As the energy depends only on the masses, the distances, and the velocities, the energies of the two systems must be identical. But the moment of momentum of the two systems is very different, for while in the one case the sum of the moments of momentum of the sun's rotation and that of the planet A, which is going in the same direction, are to be *increased* by the moment of momentum of B, the same is not the case in the other system. The moment of momentum of the sun and of A conspire, no doubt, and must be added together; but as B is revolving in the opposite direction, the moment of momentum of this planet has to be subtracted before we obtain the nett moment of momentum of the system. Hence, we perceive a remarkable difference between the two systems; for, though in each the total energy is the same, yet in the latter case the moment of momentum is smaller than in the former.

It has been pointed out that the effect of the mutual actions of the different bodies of a system is to lessen, in course of time, the total quantity of energy that they receive in the beginning, while it is not in the power of the mutual actions of the particles of the system to affect the sum total of the moment of momentum. Hence we see that, so long as the system is isolated from external interference, the tendency must ever be towards the reduction of the quantity of energy to as low a point as may be compatible with the preservation of the necessary amount of moment of momentum. The first of the two systems given are much more in conformity with this principle than the second. The moment of momentum in the former case must be nearly as large as could be obtained by any other disposition of the matter forming it, with the same amount of energy. But in the second diagram the moment of momentum is much less, though the energy is the same. It follows that the energy of this system might be largely reduced, for if accompanied by a suitable rearrangement of the planets the reduced amount of moment of momentum might be easily provided for. We thus see that this system is not one to which the evolution of a material arrangement would ultimately tend. It is, therefore, not to be expected in Nature, and we do not find it. Of course, the same would be equally true if, instead of having merely two planets, as I have here supposed for the sake of illustration, the planets were much more numerous. The operation of the causes we have been considering will show that, in the evolution of such a system, there will be a tendency for the planets to revolve in the same direction.

It is easy to see how, in the contraction of the original nebula, there must have been a strong influence to check and efface any movements antagonistic to the general direction of the rotation of the nebula. If particles revolve in a direction opposite to the current pursued by the majority of particles, there would be collisions and frictions, and these collisions and frictions will, of course, find expression in the production of equivalent quantities of heat. That heat will, in due course, be radiated away at the expense of the energy of the system, and consequently, so long as any contrary movements exist, there will be an exceptional loss of energy from this cause. Thus the energy would incessantly tend to decline. As the shrinking of the body proceeded while the moment of momentum would have to be sustained, this would incessantly tend more and more to require from all the particles a movement in the same direction.

The second concord of the planetary system, which is implied in the fact that all the planets go round in the same direction, need not therefore surprise us. It is a consequence, an inevitable consequence, of the evolution of that system from the great primæval nebula. We have seen that it would be excessively improbable that even nine or ten planets should revolve round the sun in the same direction, if the directions of their movements had been merely decided by chance. We have seen that the movements of the hosts of planets, which actually form our system, would be inconceivable, unless there were some reason for those movements. The chances against such an arrangement having arisen without some predisposing cause is so vast that, even if the chances were infinite, the case would be hardly strengthened. But once we grant that the system originated from the contraction of the primæval nebula, dynamics offers ready aid, and the difficulty vanishes. Not only do we see most excellent reasons why all the planets should revolve in the same direction; we are also provided with illustrations of similar evolutions in progress in other parts of the universe; we learn that the evolving nebula, however erratic may have been its primitive motion, whatever cross currents may have agitated it in the early phases of a possibly violent origin, will ultimately attain a rotation uniform in direction. As the evolution proceeds, the various parts of the nebula draw together to form the planets of the future system, and the planets retain the movement possessed by their component particles. Thus we see that the nebular theory not only extricates us from the difficulty of trying to explain something which seemed almost infinitely improbable, but it also shows why no other disposition of the motions than that which we actually find could be expected. The nebular theory explains to us why there is no exception to that fundamental law in the solar system which declares that the orbits of the planets shall all be followed in the same direction.

This wonderful agreement in the movements of the planets, which we have called the second concord, thus affords us striking evidence of the general truth of the nebular theory. But there is yet a third concord in the solar system which, like the other two, lends wonderful corroboration to the sublime doctrine of Kant and Laplace. This we shall consider in the next chapter.

Chapter 16

THE THIRD CONCORD

We have seen in the last chapter how the rotation of the sun beat time, as it were, for the planets, by giving to them an indication of the direction in which the revolutions round the sun should be performed, and we have observed with what marvellous unanimity the planets follow the precept thus given. We have now to consider yet another concord, which has perhaps not the great numerical strength of that last considered, but is, nevertheless, worthy of our most special attention. The earth revolves about an axis which is not very far from being perpendicular to the principal plane to which the movements of the solar system are related. From a dynamical point of view it would, of course, have been equally possible for the earth to revolve on its axis in the same direction as the rotation of the sun, or in the opposite direction. There is nothing so far as the welfare of man is concerned to make one direction of rotation preferable to the other, but, as a matter of fact, the earth does turn round in the same way as the sun turns.

Jupiter also turns on its axis, and Jupiter again, like the earth, might turn either with the sun or it might turn in the opposite direction. Here, again, we find a unanimity between the earth and Jupiter; they both turn in the same direction, and that is the direction in which the sun rotates. The same may be said of Mars, and the same may be said of Saturn. In the case of the planets Mercury and Venus we cannot speak with equal definiteness on the subject of their rotations about their axes. The circumstances of these planets are such that there are great difficulties attending the exact telescopic determination of their periods of rotation. The widest variations appear in the periods which have been assigned. It has, for instance, been believed that Venus rotates in a period not greatly differing from the period of twenty-four hours in which our earth revolves. But it has been lately supposed that the period of Venus is very much longer, and is in fact no less than seven months, which is, indeed, that of the revolution of Venus about the sun. According to this view, Venus rotates round the sun in a period equal to its revolution. If this be so, then Venus constantly turns the same face to the sun, and the movement of the planet would thus resemble the movement of the moon around the earth. As a matter of observation, the question must still be considered unsettled, though there are sound dynamical reasons for believing that the long period is much more probable than the short one. We do not now enter into this question, or into the still more difficult matter of the rotation of Mercury; it suffices to say that whichever period be adopted for either of these planets is really not material to our present argument. In both cases it has never been doubted that the direction of the rotation of the planets is the same as the direction in which Jupiter and Mars and the earth rotate, these being also the same as the direction of the solar rotation.

As to the rotations of Uranus and Neptune about their respective axes, the telescope can show us nothing. The remoteness of both these planets is such that we are unable to discern objects on their discs with the definiteness that would be required if we desired to watch their rotations. We have also no information as to the rotation of the several asteroids. No one, I think, will doubt that each of these small planets, equally with the large planets, does rotate about its axis; but it is impossible for us to say so from actual knowledge.

But undoubtedly the five old planets, Mercury, Venus, Mars, Jupiter, and Saturn, as well as the earth, all rotate in the same direction as the sun. Each planet might rotate twice as fast, or half as fast, as it does at present. They might all rotate in the opposite direction from that in which they do now, or some of them might go in one direction, and some in the other, with every variety in their diurnal periods, while the primary condition of Kepler's Laws would have still been complied with. We may also note that the *direction* in which the rotation takes place seems quite immaterial so far as the welfare of the inhabitants on these planets is concerned.

The fact that the planets and the sun have this third concord demands special attention. The chance that the earth should rotate in the same direction as the sun is, of course, expressed by one-half. It is easy to show, that out of sixty-four possible arrangements of the directions of rotation of the five planets and the earth, there would be only one in which all the movements coincided with the direction of the rotation of the sun. If, therefore, it had been by chance that the direction of these motions was determined, then Nature would have taken a course of which the probability was only one sixty-fourth. No doubt this figure is by no means so large as those which expressed the probabilities of the other planetary concords; it is, however, quite sufficient to convince us that the direction of the rotation of the planets on their axes has not been determined merely by the operation of chance.

We are to see if there is any physical agent by which the planets have been forced to turn round in the same direction. And here comes in one of those subtle points which the metaphysical genius of Kant suggested. Let us take any two planets—say, for instance, the earth and Jupiter—and let us endeavour to see what the nature of the agent must have been which has operated on these planets so as to make them both rotate in the same direction. Kant urged that there must have been some material agent working on the materials in Jupiter, and some material agent working on those of the earth, and that to produce the like effect in each planet there must have been at one time a material connection existing between that body which is now Jupiter and that body which is now the earth. In like manner Kant saw this material connection existing between the other planets and the sun, and thus he was led to see that the whole material of our solar system must once have formed a more or less continuous object. The argument is a delicate one, but it seems certainly true that in the present arrangement of the orbits it is impossible for us to conceive how, with intervals of empty space between the tracks of the planets, a common influence can have been exerted so as to give them all rotations in the same direction.

The nebular theory at once supplies the explanation of the unanimity in the rotation of the planets, just as it supplied the explanation of the unanimity in the directions of their revolutions. To explain the rotation of a planet on its axis, let us imagine that one portion of the contracting nebula has acquired exceptional density. In virtue of its superior attraction it absorbs more and more material from the adjacent parts of the nebula, and this will ultimately be consolidated into the planet, though in its initial stages this contracting matter will remain part of the nebula. We have shown that the law which decrees that the moment of momentum must remain constant will require that, after a certain advance in the contraction, all the parts of the nebula shall rotate in the same direction. Thus we find that the sun, or rather the parts of the nebula that are to form the sun, and the parts that are to form the planets are all turning round together.

At this point we may consider a geometrical principle which, though really quite simple, is not always easily understood. It has indeed presented considerable difficulty to many people. Suppose that an ordinary card is laid on a flat board, and that, with a bradawl, a hole is made through the card into the board. The hole may be at the centre, or at one of the corners, or a little way in from one of the edges, or in any other position whatever on the card. Now suppose that a postage stamp is stuck upon the card anywhere, and that the card is then moved around the bradawl. How are we to describe the motion of that postage stamp? It would certainly be revolving around the bradawl; but this motion we may consider as composed of two others. At any instant we may accurately represent the movement of the postage stamp by considering that its centre was moving in a direction perpendicular to the line joining that centre to the hole made by the bradawl, and that it also had a rotation around its centre, the period of that rotation being just the same as the time the card would take to go round the bradawl. Thus we see that the movement of the postage stamp contains at any moment a movement of translation and a movement of rotation.

Figure 55. An elongated irregular Nebula (n.g.c. 6992; in Cygnus). (Dr. W. E. Wilson, F.R.S.) *(From the Astronomical and Physical Researches at Daramona Observatory.)*

We may illustrate the case we have supposed by the movement of the moon around the earth. If the centre of the earth be considered to be at the centre of rotation the moon may be considered to be in the position of the postage stamp. As our satellite revolves, the same side of the moon is continually turned towards the earth, but this is due to the fact that the moon, at each moment, really possesses two movements, namely, a movement of translation of its centre, in a direction perpendicular to the line from the moon's centre to the earth's centre, coupled with a slow rotation of the moon round its axis.

The contracting nebula we may liken to our piece of cardboard, the stamp will represent the spot in which the nebulous material has contracted to form the planet, and the position of the bradawl is the centre of the sun. As we have seen by our illustration, the nebulous planet is endowed with a certain movement of rotation, the period of its rotation on its axis being equal to that of its revolution around the centre; and it is important also to notice that both these movements take place in the same direction.

Thus we see from the nebular theory how the primæval nebula, in the course of its contraction, originated a planet, and how that planet was also endowed with a movement of rotation; its period of rotation being originally equal to the period of rotation of the whole nebula. This explains how the planet, or rather the materials which are to form the future planet, derived from the nebula their movement of rotation, which must have been extremely slow in the beginning. As the contraction continued, the materials of the gradually growing globe drew themselves together, and tended to become separate from the surrounding nebula. At length the time arrived when the planet became sufficiently isolated from the rest of the nebula to permit the conservation of moment of momentum to be applied to it individually. Thus, though the rotation was at first excessively slow, yet, as the contraction proceeded, and as the parts of the forming planet drew themselves closer together, in consequence of their mutual attractions, it became necessary that the speed with which these parts accomplished their revolutions should be accelerated. At last, when the planet had become consolidated, and when consequently the mutual distances of the several particles constituting the planet had been reduced to but a fraction of what those distances were originally, the speed of the planet's rotation had become enormously increased. In this manner we learn how, from the very slow rotation which the nebulous material had at first, a solid planet may be made to rotate on its axis as rapidly as the planets in the solar system do to-day.

We thus find that the third concord, namely, the agreement in the directions of the planets' rotations, is a further strong corroboration of the nebular theory. The unanimity of all these various movements is the dominant characteristic of the solar system.

But this third concord, derived from the rotation of the planets, may be yet further strengthened. The movements of the satellites, which accompany so many of the planets, must also find their explanation from the primæval nebula. The circumstances of the satellites are, however, different in the different cases.

As regards the moon, the theory of its evolution is now well known, mainly by the researches of Professor George Darwin. In the moon there appear to have been causes at work of a somewhat special kind. We must just refer to what is well known with regard to the history of the moon. Here, again, we observe the importance of the principles of the conservation of moment of momentum. As the moon raises tides on the ocean surrounding the earth, and as those tides flow around the globe, they cause friction, and that friction involves, as we have so often pointed out, the loss of energy to the system. Thus, the energy of the earth-moon system must be declining, while the moment of momentum remains constant. Now there are only two sources from which the energy can be derived. One of those sources is that due to the rotation of the earth on its axis. The other is due to the moon, and consists of two parts, namely, the energy arising from the velocity of the moon in its orbit, and the energy due to the distance by which the earth is separated from the moon. As the moon's velocity depends upon its distance, we cannot view these two portions as independent. They are connected together, and we associate them into one. So that we say the total energy of the earth-moon system consists partly of that due to the rotation of the earth on its axis, and partly of that due to the revolution of the moon around the earth. It might also seem that we ought to add to this the energy due to the rotation of the moon around its own axis; but this is too inconsiderable to need attention. In the first place, the moon is so small that even if it rotated as rapidly as the earth the energy due to the rotation would not be important. Seeing, however, that the moon has for the rotation on its axis a period of between twenty-seven and twenty-eight days, its velocity of rotation is so small that, for this reason also, the energy of rotation would be inconsiderable. We are, therefore, amply justified in omitting from our present consideration the energy due to the rotation of the moon on its axis.

The energy of the earth-moon system is on the decline: the lost energy might conceivably be drawn from the rotation of the earth, or it might be drawn from the revolution of the moon, or it might be drawn from both If it were drawn from the revolution of the moon, that would imply that the moon would lose some of its speed or some of its distance, or in any case that the moon would get nearer to the earth and revolve more slowly, the speed of the earth being on this supposition unaltered. In this case, the moment of momentum of the earth would remain the same as before, while the moment of momentum of the moon would be lessened; the total moment of momentum would therefore have decreased, but this we have seen to be impossible. It therefore follows that the energy withdrawn from the earth-moon system is not to be obtained at the expense of the revolution of the moon.

The energy must therefore be obtained at the expense of the rotation of the earth on its axis. But if this be the case, the speed with which the earth rotates must be diminished; that is to say, the length of the day must be increased. And if the speed of the earth's rotation be reduced, that means that the amount of moment of momentum contributed by the earth is lessened. But the total quantity of moment of momentum must be sustained, and this can only be done by making the moon go further away and describe a larger orbit. We thus see that in consequence of the tides the length of the day must be increasing, and the moon must be gradually retreating. Thus we find that at earlier periods the moon's distance from the earth must have been less than it is at present, and the further we look back through remote periods the less do we find the distance between the earth and the moon. Thus we see that there must have been a time when the moon or the materials of the moon were in actual contact with the materials of the earth. In fact, it seems quite possible that the moon may have been a portion of the earth, broken off at some very early period, while the earth was still in a liquid state, if indeed it had condensed to even that extent. Thus the revolution of the moon round the earth is hardly to be used as an argument in favour of the nebular hypothesis. The moon is indeed a consequence of the earth's rotation.

The satellites of Mars offer conditions of a very different kind, though here, again, tidal influences have been so important, that it is perhaps questions relating to tides that are illustrated by these satellites rather than the nebular theory.

A remarkable circumstance may be noted with regard to the movements of the satellites of Mars. The inner satellite has a period of about seven and a half hours, which is not a third of the period that the planet itself takes to go round on its axis. This leads to a somewhat curious consequence. The tides raised on Mars by this inner satellite would certainly tend rather to accelerate the rotation of the planet than to retard it; for these tides must course round the planet in the direction of its rotation, but with a speed in excess of that rotation. Any tidal friction, so far as this satellite is concerned, will tend to augment the velocity of the planet's rotation, just as in the opposite case, where the moon raises tides on the earth, it is the lagging of the tides behind the movement due to the rotation that acts as a brake, and tends to check that speed. If, therefore. Mars is accelerated by this satellite, it will do more than its original share of the moment of momentum of the Martian system; it is therefore imperative that the satellite shall do less. Accordingly, we find that this satellite must go in towards the planet. No doubt this effect is much complicated by the influence of the other satellite of the same planet, but the illustration may suffice to show that if the satellites of the earth and Mars do not convey to us much direct evidence with regard to the nebular theory, this is largely because the effect of the tides has been a preponderating influence. The Martian system as we now see it has acquired its characteristic features by tidal influence, so that the more simple influences which would immediately illustrate the nebular theory have become hidden.

As to the satellites of Jupiter and Saturn, the circumstances are again quite different from those that we find in the earth and in Mars. There is little more to be said with regard to them than that everything that they present to us is consistent with the indications of the nebular theory. The evolution in each case has been a reproduction in miniature of the evolution of the solar system.

But the satellites of Uranus and Neptune present, it must be admitted, the greatest stumbling block to the acceptance of the nebular theory. Both as to the directions in which they move and as to the planes in which their orbits lie, it must be admitted that the satellites of Uranus are distinctly at variance with what the nebular theory would suggest. The consideration of this subject will be found in the next chapter.

Chapter 17

OBJECTIONS TO THE NEBULAR THEORY

No one will deny that there are points in connection with the nebular theory which present difficulties which to some seem important. We shall endeavour to estimate the significance of these difficulties in this chapter. They are certain anomalous phenomena presented by the planets Uranus and Neptune.

The satellites which attend upon the planets exhibit a general conformity with those movements of the planets themselves on which we have dwelt in Chapters 14, 15, 16. The planes in which the orbits of the satellites are contained are usually not much inclined to the plane of the ecliptic, and the directions in which the satellites revolve also agree with the general direction of the planetary movement. We find these conditions in the one satellite of the earth, in the two satellites of Mars, in the five satellites of Jupiter, in the eight or nine satellites of Saturn; but, when we come to Uranus and Neptune, the two outermost planets, we observe a striking but most instructive violation of the laws which we have found so consistently prevailing in the other parts of the solar system.

Let me first mention the special circumstances of Uranus. It is now known that this planet has four satellites. Of these, Titania and Oberon were both discovered by Sir William Herschel on January 11th, 1787. The two remaining satellites, named Ariel and Umbriel, were not discovered for more than half a century later by Mr. Lassell, on October 24th, 1851. It is, however, just possible that they were previously seen by Sir William Herschel.

The innermost of the four satellites, Ariel, accomplishes a revolution in a day and a half, Umbriel goes round in four days and three hours, Titania in eight days and seventeen hours, and Oberon in thirteen days and eleven hours. We have already mentioned how the investigations of Newcomb show that these four satellites of Uranus revolve in the same direction and in the same plane; but this plane, instead of lying in or near the ecliptic, is very nearly perpendicular thereto, the actual angle being eighty-three degrees. This is one of the features in which the satellites of Uranus are in startling disobedience to the laws which have been so rigidly observed in most other parts of the system. But there is also a second anomaly. The direction in which the satellites move, when projected on the plane of the ecliptic, is found to be opposite to the universal direction in which all the other movements in the solar system are performed. Of course the fact that the plane of the orbits of the satellites lies so nearly at right angles to the plane of the ecliptic detracts somewhat from the significance of this circumstance. If the two planes were absolutely at right angles, there would be, of course, no projection at all, and, in the actual circumstances, the moment of momentum, when projected, loses nineteen-twentieths of its amount. It follows that in the actual position of the plane the abnormal direction in which the satellites are moving is not very material.

It must be admitted that, in the position of the plane of their orbits and the direction of their movements, the satellites of Uranus are in contrast to what a hasty consideration of the nebular theory might have led us to expect. If the orbits of those satellites had all lain close to the plane of the ecliptic, and if the direction in which the satellites revolved had also conspired with that of the revolution of Uranus round the sun, and with all the other hundreds of movements which are in the same direction, there can be no doubt that we should in this place have been appealing to the satellites of Uranus as confirmatory evidence of the truth of the nebular theory. The fact that they move in a manner so totally at variance with what might have been expected cannot therefore be overlooked.

Neptune, the outermost planet of our system, presents us also with difficulties of an analogous character. So far as the orbit of Neptune itself is concerned, it agrees entirely with the general planetary convention; the inclination of that orbit to the plane of the ecliptic is no more than six degrees, and the direction in which the outermost planet revolves round the frontier of our system is not different from the directions in which all the other planets revolve. We know nothing about the axis of rotation of Neptune except that it may be reasonably presumed to be in the same plane as the movement of its satellite. On October 10th, 1846, Lassell, with the help of his great telescope, suspected the existence of a satellite to Neptune, and he announced it definitely on July 7th, 1847. We are indebted to Newcomb for a careful investigation of the orbit of this satellite. It moves in a track which is practically circular, and it requires about five days and twenty-one hours to accomplish each revolution. Its inclination to the ecliptic is not so anomalous as in the case of Uranus, the inclination being in this case not more than thirty-five degrees. This is not much greater than the inclinations of the orbits of some of the asteroids, and it might have passed without much comment had it not been for the circumstance that the direction of motion of the satellite in this track is antagonistic to all the other movements in the solar system. This is indeed a more startling fact in some respects than the movements of the satellites of Uranus, for, as we pointed out, the plane of the orbits of the satellites of Uranus is so nearly perpendicular to the plane of the ecliptic that the direction of the movement could not be held to be of much significance. The satellite of Neptune, having an orbital inclination barely more than a third of a right angle, exhibits a retrograde movement which is in some respects the most anomalous feature in the solar system.

These circumstances connected with the satellites of Uranus and Neptune have been sometimes brought forward as arguments against the nebular theory. What Laplace would have said to them we can only conjecture, for, at the time he brought out his theory, Neptune was entirely unknown, and none of the satellites of Uranus had been observed. But it has sometimes been urged that the movements of these two systems are inconsistent with the principles of the nebular theory, and that, therefore, the nebular theory must be abandoned. I have no desire to minimise the difficulties, but I think the considerations to which I now invite attention may help to lessen them even if they do not altogether remove them. I trust, at least, we may be able to show that even those anomalous movements are not incompatible with the acceptance of the account of the origin of our solar system given by the nebular theory.

The primæval nebula may be regarded as chaotic in its earliest stages; perhaps it was like the nebulous wisps. It was chaotic in the arrangement of the material of which it is formed, and in the movements of that material. Before a disorganised nebula can become evolved into a nebula with any definite form, or into anything resembling a solar system, an immense period of time must elapse, and during that time the operation of the laws of dynamics gradually impresses certain well-marked features on the nebula, and disposes it to assume an orderly form. We have explained that no matter how the nebula originated, or no matter what may have been the irregularities in its extent or distribution, and no matter how diverse may have been the agitations of its various parts, the principles of dynamics assure us that each such nebula must, for all time, stand in some special relation to a certain particular plane. The moment of momentum which the nebula has with respect to this plane, exceeds the moment of momentum that it has with respect to any other plane. We have pointed out how, notwithstanding the vicissitudes and transformations to which, in the course of illimitable ages, the nebula must submit, its moment of momentum relatively to this plane will remain absolutely unaltered. We have shown how the energy of the nebula becomes gradually exhausted. The collisions between various particles, the frictions that will necessarily arise, and the actions which we may sufficiently describe by saying that they are of a tidal character, will all result in the transformation of energy into heat. This heat is radiated away and lost, and there is a corresponding decline in the energy of the system. To preserve its moment of momentum unaltered in the course of ages, notwithstanding the continuous reduction of energy, the materials of the nebula will ever find themselves more and more approximating to the plane, and will ever find themselves more and more compelled to revolve in the same direction. If the original size of the nebula be compared with the area of the Atlantic Ocean, the condensed form which the nebula may ultimately assume may be no larger than a coral island. If the nett moment of momentum, diffused over the space as large as the ocean, has still to be preserved in the space as large as the island, we need not be surprised that the spin of the system in its condensed form is its dominating characteristic.

In the evolution of our solar system from the primæval nebula, this operation of reducing the movements to the same plane and of requiring that all the movements shall take place in the same direction, having had play for unmeasured ages, has in the main accomplished its end. All the important bodies of the system do go round in the same direction; that much, at least, has been attained. All of them also go round in planes which are nearly coincident, but, as we have already noted, they are not yet absolutely coincident. The greatest planets have, however, very nearly become reconciled, so far as the planes of their orbits are concerned, to the condition which dynamics imposes. The same is true of the rotation of the sun on its axis. That axis is inclined at an angle of eighty-three degrees to the plane of the ecliptic, so that the sun's equator would have to be shifted only through an angle no greater than seven degrees, if it were to be placed in the plane in which it should be situated, if the condition of the smallest quantity of energy for a given amount of moment of momentum was to be realised. We find a greater discrepancy in the plane of the earth's equator. This is inclined by about twenty-three degrees to the plane of the ecliptic. Here there is some energy which might yet be expended without a diminution of the amount of moment of momentum in the system; for if the earth's axis were to be made perpendicular to the plane of the ecliptic, then the velocity of rotation of the earth about its axis might undergo a corresponding abatement, and yet keep up the requisite moment of momentum. We thus see that even with the older planets the conditions which would be enforced, if the moment of momentum was to be sustained with the least quantity of energy, are not absolutely complied with; which simply means that there has not yet been time enough for our system to arrive at the perfect state, to which it must be approximating.

If we have found that in the rotations of the earth and of the sun, and in the revolutions of the planets round the sun, the conditions ultimately aimed at have not yet been reached, why should we feel surprised that in the outer planets of our system, Uranus and Neptune, the conditions which evolution tends to produce have not yet been fully attained? That the operation of the conservation of moment of momentum is in progress in the internal economy of the Uranian system, we have already had occasion to explain in Chapter 11. The fact which Newcomb demonstrated, that the four satellites revolve in the same plane, can only be accounted for by the supposition that in that system the conservation of moment of momentum, with declining energy, has gradually imposed this condition on the system belonging to Uranus. With reference to the position of the plane of the satellites, in the case of Uranus and Neptune, we would say, that though at present their arrangement appears anomalous, it will probably not always remain so. The fact that the satellites of Uranus are in a plane nearly perpendicular to the plane of the ecliptic really implies that there is a certain amount of energy still disposable in our system, if by readjustment of the plane of the Uranian satellites the necessary moment of momentum in the system is still preserved.

The laws of dynamics tell us that the orbits of planets must be gradually, if with excessive slowness, tending still further to the same plane. In this process energy can be expended by the system, while the moment of momentum is unabated. We can at least suggest what seems to be at this moment in progress in the system belonging to Uranus. It will readily be admitted that there may be a difficulty in seeing how the movement of a planet, which is going in the wrong direction, could be stopped and turned into the right direction. But we need not suppose that so violent a change as this would imply is to be expected in our system. We are quite accustomed to find the planes of the orbits of all planets in gradual movement. The plane containing the orbits of the four satellites of Uranus is at this moment probably moving gradually upwards. It will in due course become actually at right angles to the ecliptic, and we may then reasonably assume that it will advance further in the same direction. At the moment the right angle is passed, this continuous movement will have the effect of changing the directions of the satellites' movement from retrograde to direct. The present anomaly will then tend to become evanescent, for, as the exhaustion of the energy continues, the planes of the satellites of Uranus will gradually come into conformity with the plane of the ecliptic.

Figure 56. Two-branched Spiral (n.g.c. 7479; in Pegasus).
(Lick Observatory.)

We make no doubt that there may be a similar explanation of the movements of the satellite of Neptune. The inclination of the plane of the orbit of the satellite to the ecliptic is probably now increasing. It will ultimately come to be at right angles thereto, and then the next advance of the plane will convert, by a continuous action, the retrograde motion of the satellite, at present so disconcerting, into a direct motion. The change of the plane will still continue until it, too, may ultimately coalesce with the ecliptic.

The fact appears to be, that though an enormous quantity of energy must have been lost by radiation from our system during the illimitable ages through which the evolution has been running its course, the disposable energy is not yet quite exhausted. There are certain adjustments in our system which may still be made and which will allow of yet further radiation of energy, while still preserving sufficient to keep up the necessary moment of momentum. It seems obvious that the system is tending towards a condition in which the planes of all the orbits shall be coincident, and in which all the directions shall be absolutely unanimous. If we were at once to alter the system by moving all the orbits into the plane of the ecliptic, but making no change in the dimensions of those orbits, or the velocities concerned; if we were also to adjust the rotations of the earth, as well as of the other planets, so that all the axes of rotation should be perpendicular to the plane of the ecliptic; if we were to turn the plane of the satellites of Uranus through that angle of 97°, which would suffice at the same time to bring it into coincidence with the ecliptic, and lay the movements of the satellites in the right direction; if we were also to turn the orbit of the satellite of Neptune through 145°, thus bringing that orbit to coincide with the plane of the ecliptic, in such a manner that the direction of the movement of the satellite of Neptune conspired with all the other movements of the system, then this rearrangement of the system would increase the moment of momentum, while the quantity of energy was not altered. But this is the same thing as saying that some energy yet remains to be disposed of, while the system still preserves the requisite moment of momentum.

The conclusion we come to may be thus expressed: the movements of the satellites of Uranus and Neptune do not disprove the nebular hypothesis. They rather illustrate the fact that the great evolution which has wrought the solar system into form has not yet finished its work; it is still in progress. The work is very nearly done, and when that work shall have been completed, the satellites of Uranus and Neptune will no longer be dissociated from the general concord.

Chapter 18

THE BEGINNING OF THE NEBULA

Whatever may have been the antiquity of the actual elements that formed the primæval nebula from which the solar system has been evolved, the nebula itself has certainly not been of infinite duration. The question then arises as to what has been the origin of the nebula as such, or rather by what agency the material from which the nebula was formed underwent so radical a transformation from its previous condition as to be changed into that glowing object which we have considered so frequently in this book. We have to explain how, by the operation of natural causes, a dark body can be transformed into a glowing nebula.

Let us first estimate what the quantity of energy in that system is. The sun has been pouring forth heat for inimitable ages, and will doubtless continue to pour forth heat for millions of years to come. But the destiny which awaits the sun, though it may be protracted, yet cannot be averted. The sun will go on pouring forth its heat and gradually shrinking. The time will come at last when the radius of the sun will have appreciably decreased, and when once it has assumed a density corresponding to a solid state its history as a radiant globe will be approaching its close. A period of insignificant extent, a century or less, will then suffice for that solid globe to cool down so as to be no longer an efficient source of light and heat. We shall assume that when the sun has ultimately become solid and cold, and when it is no longer the life and light of our system, it will have attained a mean density of 21.5, which we have chosen because that is the density of platinum, the heaviest substance known. In all probability the solar density will never become so great as this, but to include the most extreme case in our argument I am making the assumption in the form stated. We are now to estimate what will have been the total energy that the sun has radiated from the moment when as an indefinitely great nebula it first began to radiate at all, down to that moment in the future when, having shrunk to the density of platinum, and having parted with all its heat, the solar radiation is at an end.

In the beginning of the evolutionary history the sun was a nebula, which we have supposed to extend in every direction to an indefinitely great distance. The system has resulted from the contraction of that nebula, and the energy liberated in that contraction has supplied the sun's radiation. We calculate (*see* Appendix) the energy that would be given out in the contraction of a nebula whose materials were originally at infinity, and which ultimately coalesced to form a cold, solid globe of the density of platinum, and as heavy as the sun. There is no object in attempting to express this quantity of energy in foot-pounds—the figures would convey no distinct impression—we shall employ the coal-unit explained in Chapter 6. We imagine a globe of coal the weight of the sun; then, if that globe of coal were adequately supplied with oxygen, it would, on combustion, give out a certain amount of heat, which is a convenient unit for our measurements. It is demonstrated that the quantity of energy given out by the contraction of the nebula from infinity, to this globe of the density of platinum, would be about equal to the quantity of energy which would be produced by the combustion of 8,300 globes of coal as heavy as the sun, an adequate contribution of oxygen being supposed to be supplied. This expresses the original endowment of energy in the solar system, or rather a major limit to that endowment; it shows that the solar system can never have developed more energy by contraction than that which could be produced by the combustion of 8,300 globes of coal as heavy as the sun. We may mention that of this great endowment of energy an amount which is rather less than half (3,400) has been already expended, so that rather more than half of the sun's career as a radiant globe may yet have to be run.

We can also express the total energy of the solar system in a different manner. We shall consider what must be the velocity of the sun, so that the energy that it will possess, in virtue of that velocity, shall be equal to the energy which could be produced by the combustion of 8,300 globes of coal of the same weight. This calculation is very much simplified by making use of a principle which we have already stated and applied in Chapter 5. We have shown that if a piece of coal be animated with a velocity of five miles a second, the energy it possesses in virtue of that motion is equal to the energy produced by the coal in the act of combustion. If a body were moving at the rate of, let us say, 100 miles a second—its speed being then twenty times as great as the particular speed just mentioned—its energy, which depends on the square of the velocity, would be 400 times as much as would be produced by the burning of a piece of coal equal to it in weight. We can easily calculate that if the sun were moving at a speed of 460 miles a second, it would possess, in virtue of its motion, as much energy as would be generated by the contraction of the primæval nebula from infinity down to a globe of the density of platinum.

It is thus easy to form a supposition as to how the nebula constituting our solar system may have come into being; most probably it originated in this way. Let us suppose that two masses, either dark or bright, either hot or of the temperature of space, or the temperature of frozen air, were moving with speeds of 460 miles a second. No doubt the velocities we are here postulating are very high velocities, but they are not unprecedentedly high. We know of stars which at this present moment move quite as fast, so that there is nothing unreasonable in our supposition so far as the velocities are concerned. Let us suppose that each of these bodies had a mass which is half that of our present solar system. If these two bodies dashed into collision, when moving from opposite directions, the effect of the blow would be to transform the energy into heat. That heat would be so great that it would be sufficient not alone to render these globes red-hot and white-hot, but even to fuse them—nay, further, to drive them into vapour, even to a vapour which might expand to an enormously great distance. In other words, it is quite conceivable that a collision of two such masses as we have here supposed might be adequate to the formation of a nebula such as that one which in the lapse of indefinite ages has shaped itself into the solar system.

Before the collision, which resulted in the formation of the nebula, each of these bodies, or rather their centres of gravity, would be moving in what may be regarded for the moment as straight lines, and a plane through those two straight lines will be a plane which forever afterwards will stand in important relation to the system. It will be, in fact, that principal plane of which we have so often spoken.

As those two bodies met they would possess a certain moment of momentum, and this moment of momentum would remain forever unaltered, no matter what may be the future vicissitudes of the system.

For the sake of simplicity in describing what has occurred, we have spoken as if the two bodies were of equal mass, and, moving with equal velocities from opposite points of the heavens, dashed into collision. But what actually happens cannot have been quite so symmetrical. There is one feature in the solar system which absolutely proves that the collision cannot have taken place precisely in the way we have laid down. If it had happened that two equal masses had been hurled into collision with equal velocities from precisely opposite directions, then there could have been no resultant moment of momentum. From the principle of the conservation of moment of momentum, we can see that, if absent in the beginning, it could never originate later. As, however, we have a large moment of momentum in the movements of the planets and the sun, it is certain that the collision cannot have taken place in a manner quite so simple.

The probabilities of the case show that it is almost infinitely unlikely that two bodies of equal dimensions, and moving with equal velocities in opposite directions, should come squarely into collision. It would be much more likely that the bodies should be not of the same size, not moving with the same velocity, and should collide partially rather than squarely. The collision may have been, in fact, little more than a graze. The probabilities of the case are such as to show that what actually occurred was a collision between two unequal masses, which were moving in directions inclined to each other and with different velocities. It is easy to show that, granted sufficiently great velocities, an impact which fell far short of direct collision might still produce enough heat to transform the whole solar system into vapour.

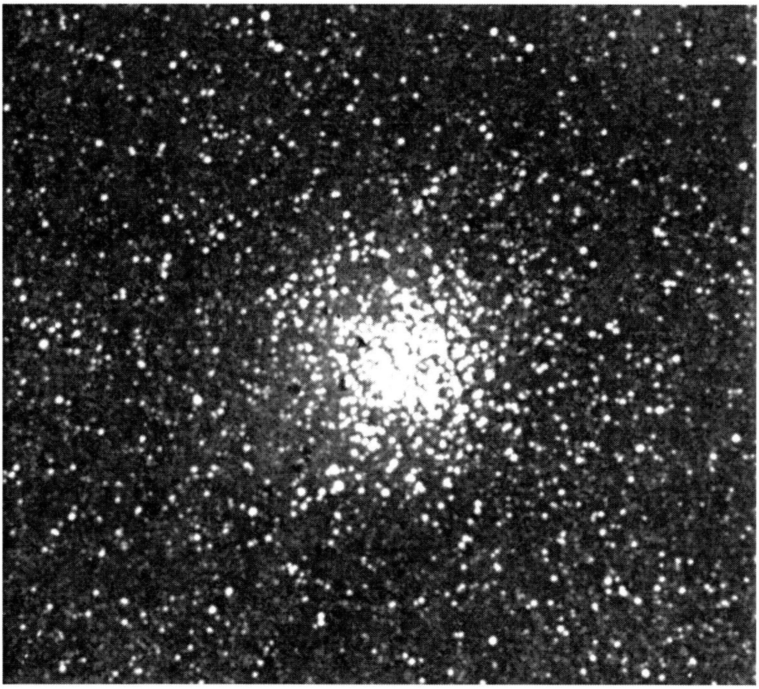

Figure 57. Cluster with Stars of 17th Magnitude (n.g.c. 6705; in Antinous). *(Photographed by Dr. Isaac Roberts, F.R.S.)*

The circumstances which would naturally accompany so transcendent an incident will also go far to account for a difficulty which has been often felt with regard to the evolution of the system from a nebula. Were such a collision to take place we should certainly not expect that the resulting nebulous mass, the product of a shock of such stupendous violence, would be a homogeneous or symmetrical object. Portions of the colliding body would become more highly heated than others; portions of the bodies would not be so completely transformed into vapour as would other parts. There would thus be differences in the nebula at the different parts of its mass. This non-homogeneity would be connected with the formation and growth of planets in the different parts of the nebula.

There is another circumstance connected with the movement of the sun which should here be mentioned. It is well known that the sun has a velocity which carries it on through space at the rate of half a million miles a day. In this movement the whole solar system, of course, participates. This movement of translation of our system must also be a result of the movements of the two original colliding masses. These two masses imparted to the system, which resulted from their union, both the lineal velocity with which it advances through space, and also that moment of momentum which is of such vast importance in the theory.

A consideration of the probabilities of the case make it quite certain that the celestial bodies we see are as nothing compared with the dark bodies we do not see. The stars we see are moving, and the natural assumption is that the dark objects with which the heavens teem are also in motion. We shall, under these conditions, not feel any insuperable difficulty in the supposition that collisions between different bodies in the heavens may have taken place from time to time. We remember that these bodies are moving in all directions, and at extremely high velocities. We are quite willing to grant the excessive improbability that any two bodies particularly specified should come into collision. Within view of our telescopes we have, however, a hundred millions of stars, and if we multiply that figure even by millions, it will still, we may well suppose, not be too large to express the number of bodies which, though contained within the region of space ranged over by our telescopes, are still totally invisible. In these circumstances, we may admit that occasional collisions are not impossible. Please note the strength which the argument derives from the enormous increase in our estimate of the number of bodies, when we include the dark objects as well as the stars. If we were asked whether it would ever be possible for two bright stars to come into collision, we might well hesitate about the answer. We know, of course, that the stars have proper motions; we know, too, that the stars, in this respect unlike the planets, have no definite directions of movement under the control of a supreme co-ordinating attraction. Some stars move to the right, and some to the left, some one way and some another; but even still, notwithstanding their great number, the extent of space is such that the stars keep widely apart, and thus collisions can hardly be expected to take place, unless perhaps in a cluster. We have no reason to think that a collision between two actual bright stars was the origin of the primæval nebula of our system. But when we reflect that the stars, properly so called, are but the visible members of an enormously greater host of objects, then the possibilities of occasional collision between a pair of these incomparably more abundant dark bodies seems to merit our close attention. We are not by any means claiming that such collisions occur frequently. But what we do say is, that if, as we believe, these bodies are to be reckoned in many millions of millions, then it does sometimes happen that two of them, moving about in space, will approach together sufficiently to give rise to a collision. It was from some such collision that we believe the nebula took its rise from which the solar system originated.

We have the best reason for knowing that celestial collisions do sometimes occur. It will be in the recollection of the readers of this chapter that in February, 1901, the astronomical world was startled by the announcement of the outbreak of a new star in Perseus. A photograph of that part of the heavens had been taken a few days before. There were the ordinary stars, such as existed from time immemorial, and such as have been represented on the numerous maps in which the stars are faithfully set down. But, on February 22nd, Dr. Anderson, already famous by similar discoveries, noticed that the constellation of Perseus contained a star which he had not seen before. Instantly the astronomical world was apprised by telegraph that a new star had appeared in Perseus, and forthwith most diligent attention was paid to its observation. Photographs then obtained show the stars that had been seen there before, with the addition of the new star that had suddenly come into view. For a few nights after its discovery the object increased in lustre, until it attained a brightness as great as that of Capella or Vega. But in this state it did not long remain. This brilliant object began to wane. Presently it could not be classed as a star of the first magnitude, nor yet of the second, and then it ran down until a little below the third, and even below the fourth. In the subsequent decline of the star there were several curious oscillations. On one night the star might be seen, the next night it would be hardly discerned, while the night after it had again risen considerably. But, notwithstanding such temporary rallies, the brightness, on the whole, declined, until at last the star dwindled to the dimensions of a small point of light, scarcely distinguishable with the naked eye. The decline was apparently not so rapid as the increase, but nevertheless from the first moment of its appearance to the last was not longer than a few weeks.

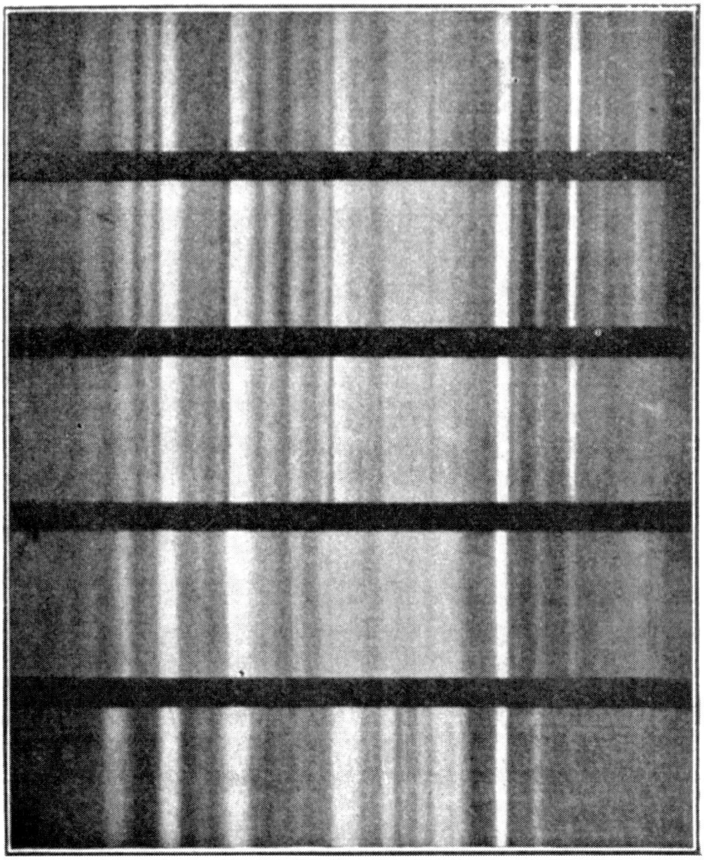

Figure 58. Spectrum of Nova Persei (1901).
(Photographed with the 40 in. Yerkes Telescope by Mr. Ferdinand Ellerman.)

This new star in Perseus established, in one sense, a record. For the star was brighter than any new star which had been noticed since the days of accurate astronomical observations. Not indeed for three centuries had a star of such lustre sprung into existence. But a temporary star, such as this was, has been by no means an infrequent occurrence. Many such have been recorded. Those who have been acquainted with astronomical matters for thirty years will recollect four or five such stars. In each of them the general character was somewhat the same. There was a sudden outbreak, and then a gradual decline. The questions have sometimes arisen as to whether the outbreak of such an object is really the temporary exaltation of a star which was previously visible, or whether it ought not to be regarded as the creation of a totally new star. In some cases it does seem possible that a new star may have been partly, at all events, due to a large increase of brightness of some star which had been known before. In the case of Nova Persei, however, we have the best authority that this is not the case. Professor Pickering, the distinguished astronomer of Harvard College Observatory, happened to photograph the region in which Nova Persei appeared a few days before the outbreak took place. He tells us that there is not the least indication on his photograph of the presence of a star in that region.

The spectrum of Nova Persei, in an instrument of sufficient power, appeared a truly magnificent object. Like other stellar spectra, it displayed the long line of light marked with the hues of the rainbow, but it was unlike the spectra of ordinary stars in respect of the enormous enhancements of the brightness at various parts of this spectrum. For instance, at one end of the long coloured band a brilliant ruby line glowed with a lustre that would at once attract attention, and demonstrated that the object under view must be something totally different from ordinary stars. This superb feature is one of the lines of hydrogen. The presence of that line showed that m the source from which the light came there must have been a remarkable outbreak of incandescent hydrogen gas. At various points along the spectrum there were other beautiful bright lines which, in each case, must have been due to glowing gas. Here we have the evidence of the spectrum telling us in unmistakable language that there were features in this star wholly unlike the features found in any ordinary star. It is impossible to dissociate these facts from the history of the star. Much of what we have said with regard to the spectrum of Nova Persei might be repeated with regard to the spectrum of the other temporary stars which, from time to time, have burst forth. In each case the spectrum characteristic of an ordinary star is present, but superadded to it are bright lines which indicate that some great convulsion has taken place, a convulsion by which vast volumes of gas have been rendered incandescent. We show the spectrum of Nova Persei on five dates, from February 27th to March 28th, 1901. These photographs were taken by Mr. Ferdinand Ellerman with the great telescope of the Yerkes Observatory. They show in the clearest manner the bright lines indicating the incandescent gases.

We have pointed out the high probability that among the millions and millions of bodies in the universe it may now and then happen that a collision takes place. Have we not also explained how the heat generated in virtue of such a collision might be sufficient, and, indeed, much more than sufficient, to raise the masses of the two colliding bodies to a state of vivid incandescence? A collision affords the simplest explanation of the sudden outbreak of the star, and also accounts for the remarkable spectrum which the star exhibits.

Chapter 19

CONCLUSION

It is not difficult to show that the nebular theory occupies a unique position among other speculations of the human intellect. It is so comprehensive that almost every conceivable topic will bear some relation to it. Perhaps I may venture to give a rather curious illustration of this fact, which was told me many years ago by one who attended a course of lectures by an eminent Professor in the medical faculty at, let us say, Vienna. The subject of the course was the no doubt highly important, but possibly not generally interesting, subject of "inflammation." I think I am right in saying that the course had to last for six months, because the subject was to be treated with characteristic breadth and profundity. At all events, I distinctly remember that the learned Professor commenced his long series of professional discourses with an account of the nebular theory, and from that starting point he gradually evolved the sequence of events which ultimately culminated in—inflammation!

It may be remembered that in the year 1880, Professor Huxley delivered at the Royal Institution a famous lecture which he termed "The Coming of Age of the Origin of Species." Among the many remarkable and forcible illustrations which this lecture contained, I recall one which brought before the audience, in the most convincing manner, the truth of the great Darwinian Theory of Evolution. Huxley pointed out how the discoveries in Biology, during the twenty-one years which immediately succeeded the publication of the "Origin of Species," had been so numerous and so important, and had a bearing so remarkable on the great evolutionary theory, that even if the Darwinian Theory had not been formed to explain the facts of Nature, as they were known at the time when Darwin published his immortal book, the same theory would have had to be formed, were it only to explain the additional facts which had come to light since the great theory itself had been first given to the world.

I believe we may use similar language with regard to the nebular theory and its great founders, Kant, Laplace, and Herschel. If the facts which were known to these philosophers led them to adopt in one form or another that view of the Origin of the Universe which the nebular theory suggests, how stands the theory now in the light of the additional facts that have been since disclosed? If we merely took the discoveries which have been made since the last of the three great philosophers passed away, it might well be maintained that a nebular theory would be demanded to account for the facts brought to light, in the interval.

The argument on which the nebular theory of the solar system is founded has other parallels with that wonderful doctrine of Natural Selection by which Darwin revealed the history of life on our globe. It not unfrequently happens that an animal has in its organisation some rudiments of a structure which is obviously of no use to the animal in his present mode of life, and would be unintelligible if we supposed the animal to have been created as he is. A curious instance of a rudimentary structure is furnished in the apteryx, the famous wingless bird which still lives in New Zealand.

The arrival of civilisation in New Zealand seems likely to be accompanied with fatal results, so far as the unfortunate apteryx is concerned. Weasels and other fierce enemies have been introduced, with which this quaint bird of antiquity is unable to cope. The apteryx is defenceless against such foes. Nature had not endowed it with weapons wherewith to fight, for it had, apparently, no serious adversaries until these importations appeared in its island home. Unlike the ostrich, the apteryx has neither strength to fight his enemies, nor speed to run away from them, though, like the ostrich, it has no wings for flight; indeed, the apteryx has no wings at all. As its name signifies the apteryx is the wingless bird. Living specimens are still to be seen in the Zoological Gardens. The special point to notice is that, though he has no wings whatever, still there are small rudimentary wing-bones which can be easily seen. You need not be afraid to put your hand on the apteryx, and feel the puny little remnants of wings.

Figure 59. The Apteryx: A Wingless Bird of New Zealand.

If, having seen the bird in the Zoological Gardens, you go to the Natural History Museum, you will there find a skeleton of the apteryx. Look near the ribs in the photograph, and there you will see those poor little wing-bones—wing-bones where there never was a wing. From our present point of view these wings are, however, more interesting and instructive than the most perfect wings of an eagle or a carrier-pigeon. Those wings in the apteryx may be incapable of flight, but they are full of instruction to the lover of Nature. As it is certain that they are absolutely of no use whatever to the bird, we may well ask, why are they there? They are not there to give assistance to the bird in his struggle for life; they cannot help him to escape from his enemies or to procure his food; they cannot help him to tend and nurture the young one which is hatched from the egg; they can help him in no way. The explanation of those ineffectual wings is historical. Those bones are present in the apteryx simply because that bird has come down by a long line of descent from birds which were endowed with genuine wings, with wings which enabled them to fly like rooks or partridges.

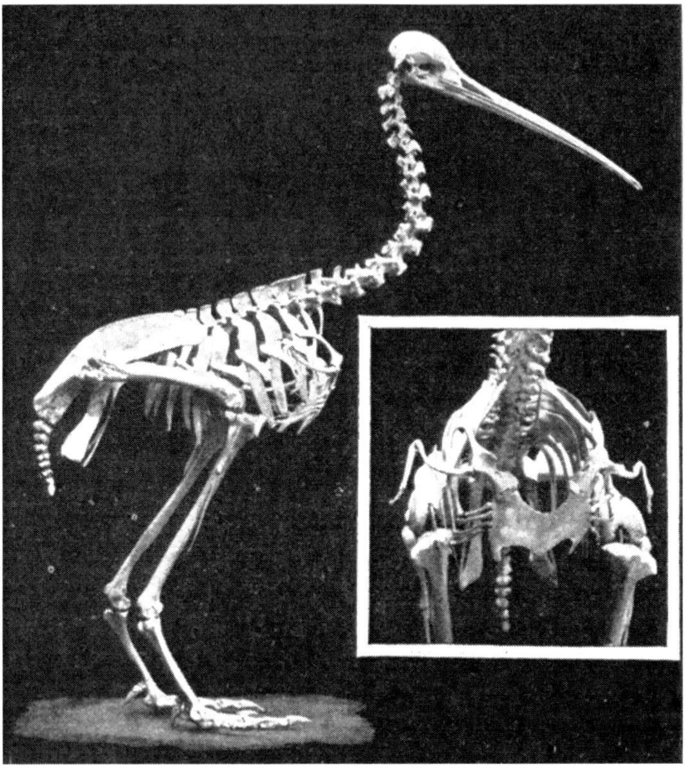

Figure 60. Skeleton of the Apteryx, showing Rudimentary Wings.

But if this be the explanation, how has it come to pass that the wings have dwindled to useless little bones? We cannot of course feel certain of the reason, but it seems possible to make surmises. In early times winged birds flew over the sea into New Zealand, and found it a country of abundance, as many other immigrants have done in later times. It may have been that the food in New Zealand was so plentiful thatwants of the birds could be readily supplied, without the necessity for ranging over large tracts. It may have been that the newly arrived birds found that they had few or no enemies in New Zealand, from which flight would be necessary as a means of escape. It may possibly have been both causes together, and doubtless there must have been other causes as well. The fact is, however, certain, that in the course of long generations this bird gradually lost the power of flight. Natural selection decrees that an organ which has ceased to serve a useful purpose shall deteriorate in the course of generations. If the wings had become needless in the search for food, unnecessary for escape from enemies, and useless for protection of its young, they would certainly tend towards disappearance. The organism finds it uneconomical to maintain the nutrition of a structure which discharges no useful end. The wings, in such circumstances, would be an encumbrance rather than an aid, and so we may readily conjecture that, in accordance with this well-known principle, the wings gradually declined, until they ceased to be useful organs, so that now merely a few rudimentary bones remain to show that the bird's ancestors had once been as other birds. Whatever may have been the cause, it seems certain that in the course of thousands of years, or it may be in scores of thousands of years, these birds lost the power of flight; thus they gradually ceased to have wings, and these little bones are all that now remain to render it almost certain that, if we could learn what this bird's ancestry has been, we should find that it was descended from a bird which had useful wings and vigorous flight. Whenever we find an organ which is obviously rudimentary, or of no use to its possessor in its present form, Darwin has taught us to look for an historical explanation. Let us see if we cannot apply this principle to the illustration of the nebular theory.

We liken the internal heat of the earth to the rudimentary wing-bones of the apteryx. In each case we find a survival devoid of much significance, unless in regard to its historical interpretation. But that historical significance can hardly be over-estimated. Unimportant as the wing-bones may be, they admit of explanation only on the supposition that the apteryx was descended from a winged ancestor. Unimportant as the internal heat, still lingering in our globe, may seem, it admits of explanation only on the supposition that the earth has had the origin which the nebular theory suggests.

That the earth's beginning has been substantially in accordance with the great Nebular Theory is, I believe, now very generally admitted. But the only authority I shall cite in illustration of this final statement is the Lady Psyche, who commences her exquisite address to her "patient range of pupils" with the words:—

> "This world was once a fluid haze of light,
> Till toward the centre set the starry tides,
> And eddied into suns, that wheeling, cast
> The planets;"

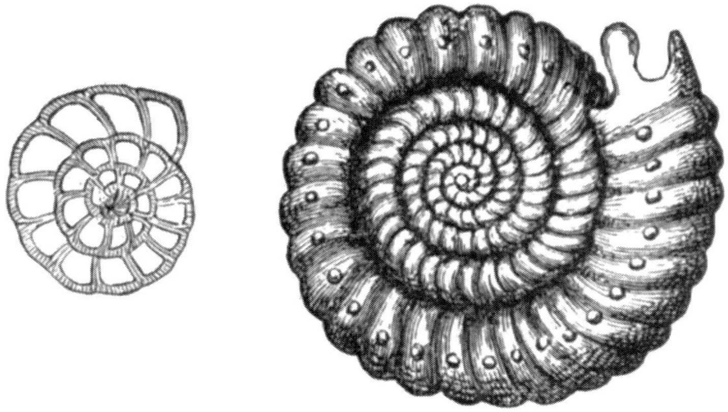

Figure 61. Foraminifer. and Nautilus. Spirals in other Departments of Nature.

APPENDICES

I. ON THE HEAT GIVEN OUT IN THE CONTRACTION OF THE NEBULA

§ 1. Fundamental Theorems in the Attraction of Gravitation

The first theorem to be proved is as follows:—

> The attraction of a thin homogeneous spherical shell on any point in its interior vanishes.

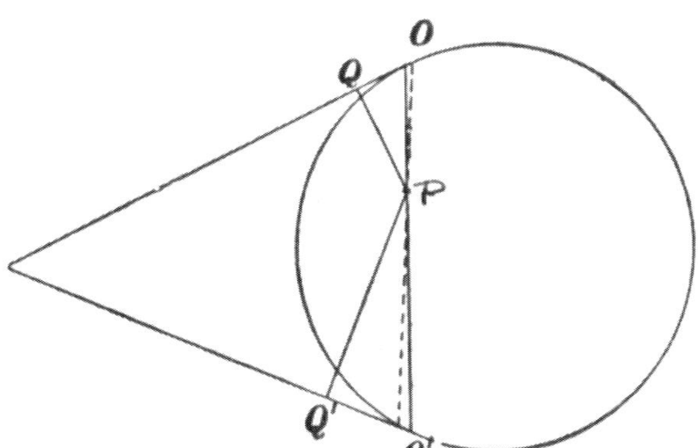

Figure 62.

Take any point P within the sphere. Let this be the vertex of a cone produced both ways, but with a very small vertical angle, so that the small areas S and S´, in which the two parts of the cone cut the sphere, may be regarded as planes. Draw the tangent planes at S and S´. Let the plane of the paper pass through P and be perpendicular to both these tangent planes. Let O P O´ be one of the generators of the cone, and let fall P Q perpendicular to the tangent plane at O, and P Q´ perpendicular to the tangent plane at O´. The volume of the cone with the vertex at P and the base S is 370⅓ P Q × S, and the other part of the cone has the volume ⅓ P Q´ × S´.

As the vertical angles of the cones are small, their volumes will, in the limit, be in the ratio of $O\,P^3$ to $O'\,P^3$, and accordingly $\frac{1}{3}\,P\,Q \cdot S \div \frac{1}{3}\,P\,Q' \cdot S' = P\,O^3 \div O'\,P^3$. But from the figure $P\,Q \div P\,Q' = P\,O \div P\,O'$, and hence $S \div O\,P^2 = S' \div O'\,P^2$.

As the shell is uniform, the masses of the parts cut out by the cones are respectively proportional to S and S´. Hence we see that the attractions of S and S´ on P will neutralise. The same must be true for every such cone through P, and accordingly the total attraction of the shell on a particle inside is zero.

The second fundamental theorem is as follows:—

> A thin spherical homogeneous shell produces the same attraction at an external point as if its entire mass were concentrated at the centre of the sphere.

This is another famous theorem due to Newton. He gives a beautiful geometrical proof in Section XII. of the first book of the “Principia.” We shall here take it for granted, and we shall consequently assume that—

> The attraction by the law of gravitation of a homogeneous sphere on an external point is the same as if the entire mass of the sphere were concentrated at its centre.

§ 2. On the Energy between Two Attracting Masses

Let m and m' be two attracting bodies supposed to be small in comparison with their distance x. Let the force between them be $\varepsilon\, m\, m' \div x^2$ when ε is the force between two unit masses at unit distance. It is required to find the energy necessary to separate them to infinity, it being supposed that they start from an initial distance a. The energy required is obtained by integrating between the limits infinity and a, and is consequently $\varepsilon\, m\, m' \div a$.

§ 3. On the Energy Given Out in the Contraction of the Nebula

We assume that the nebula is contracting symmetrically, so that at any moment it is a homogeneous sphere. We shall consider the shell which lies between the two spheres of radii, $r + dr$ and r respectively.

Let M´ be the mass of the nebula contained within the sphere of radius r, and let dM´ be the mass of the shell just defined. Then it follows from § 1 that the condensation of the shell will have been effected by the attraction of the mass M´ solely. The exterior parts of the nebula can have had no effect, for the outer part has always been in symmetrical spherical shells exterior to dM´, and the attraction of these is zero. We see from § 2 that the contraction of dM´ from infinity, until it forms a shell with radius r, represents a quantity of energy,

$$(\varepsilon\, M' dM')/r;$$

for it is obvious that the energy involved in the contraction of the whole shell is the sum of the energies corresponding to its several parts.

If M be the total mass and a the radius of the nebula always supposed homogeneous

$$M' = M\,(r^3/a^3),$$

and therefore

$$dM' = 3\,M\,(r^2/a^3)\,dr.$$

Hence the work done in the contraction is

$$(\varepsilon/r)\,M\,(r^3/a^3)\cdot 3\,M\,(r^2/a^3)\,dr = (3\,\varepsilon/a^6)\,M^2\,r^4\,dr.$$

Integrating, therefore, the total work of contraction is

$$\tfrac{3}{5}\,(\varepsilon\,M^2/a)$$

At the present moment a mass of 1 lb. at the surface of the sun would weigh 27 lbs. if tested by a spring balance. Hence

$$\varepsilon\,M/a^2 = 27.$$

With this substitution we find the expression for the foot-pounds of work corresponding to the contraction of the nebula from infinity to a sphere of radius *a* to be,

$$\tfrac{3}{5}\cdot 27\,a\,M = 16\,a\,M \text{ very nearly.}$$

Hence we have the following fundamental theorem due to Helmholtz, which is the basis of the theory of sun heat.

> If the sun he regarded as a homogeneous sphere of mass M pounds and radius a feet, then the foot-pounds of energy rendered available for sun heat by the contraction of the solar material from, an infinite distance is 16 a M.

§ 4. Evaluation of the Sun Heat Given out in Contraction

The number of foot-pounds of work given out in the contraction from infinity is 16 *a* M. As 772 foot-pounds are equal to ne unit of heat, i.e., to the quantity of heat necessary to raise 1 lb. of water 1° Fahrenheit, we see that 772 M is the work required to raise a mass of water equal to the mass of the sun through 1° Fahrenheit. Hence the number of globes of water, each equal to the sun in mass, which would be raised 1° Fahrenheit by the total heat arising from the contraction, is

$$(16\,a)/772,$$

but a, the radius of the sun in feet, is 2,280,000,000, and hence we have the following theorem:—

> The energy liberated in the contraction of the sun from infinity to its present dimensions would, if turned into heat, suffice to raise 47,000,000 globes of water, each having the same mass as the sun, through 1° Fahr.

It is found by experiment that 1 lb. of good coal may develop 14,000 units of heat, and is therefore equivalent to 14,000 × 772 foot-pounds of work. A mass of coal equal to the sun would therefore (granted oxygen enough) be equivalent to 14,000 × 772 × M foot-pounds of work. But we have

$$\frac{(16\,a\,\mathrm{M})}{14{,}000 \times 772 \times \mathrm{M}} = \frac{16 \times 2{,}280{,}000{,}000}{14{,}000 \times 772} = 3{,}400.$$

Hence we see that

> The energy liberated in the contraction of the sun from infinity to its present dimensions, is as great as could be produced by the combustion of 3,400 globes of coal, each as heavy as the sun.

We may speak of 3,400 in this case as the coal equivalent.

§ 5. On the Further Contraction of the Sun and the Heat That May Thus be Given Out

Let us suppose the sun contracts to the radius r, and then, as already proved, § 3, the energy it gives out is

$$\tfrac{3}{5} (\varepsilon\, M^2)/r,$$

but we have

$$\varepsilon\, M/a^2 = 27,$$

whence on contraction to the radius r the total energy given out from the commencement is

$$16\, M\, (a^2/r)$$

The average density of the sun at present is 1.4. Let us suppose it condenses until it has a density ρ.

$$r^3 \div a^3 = 1.4 \div \rho,$$

whence the energy becomes

$$14\, a\, M \cdot \sqrt[3]{\rho};$$

but the coal equivalent of 16 a M has been found in § 4 to be 3,400, and hence the coal equivalent in this case is

$$3{,}000\, \sqrt[3]{\rho}.$$

If we take ρ to be the density of platinum (21.5), we get a coal equivalent 8,300. This, therefore, seems to represent a major limit to the quantity of heat which can be obtained from the condensation of the nebula from infinity into a sun of the utmost density.

§ 6. On the Present Emission of Sun Heat

According to Scheiner, "Strahlung und Temperatur der Sonne, Leipzig, 1899," the value of the solar constant, i.e., the number of cubic centimetres of water which would be raised 1° Centigrade by the quantity of sun heat which, if there were no atmospheric absorption, would fall perpendicularly on a square centimetre, at the earth's mean distance from the sun, is between 3.5 and 4.0. If we take the mean value, we have (translated into British units), the following statement:—

> If at a point in space, distant from the sun by the earth's mean distance, one square foot was exposed perpendicularly to the solar rays, then the sun heat that would fall upon it in one minute would raise one pound of water 14° Fahr.

This shows that the solar energy emitted daily amounts to

$$700{,}000{,}000{,}000 \times 4\,\pi\,a^2 \text{ foot-pounds.}$$

§ 7. On the Daily Contraction of the Sun Necessary to Supply the Present Expenditure of Heat

We have seen that at the radius r the energy is

$$16\,\mathrm{M}\,(a^2/r).$$

Hence for a change dr it is

$-16\ M\ (a^2/r^2)\ dr.$

At its present size, accordingly, the energy given out by a shrinkage *dr* is

$16\ M\ dr.$

One cubic foot of the sun averages 87 pounds, so that

$M = \frac{4}{3}\pi a^3 \times 87$

$16\ M\ dr = 464 \times 4\ \pi\ a^3\ dr.$

We have to equate this to the expression in the last article, and we get

$dr = 700{,}000{,}000{,}000/(464\ a) = .65.$

This is the shrinkage of the sun's radius expressed in feet. Hence the daily reduction of the sun's *diameter* is 16 inches.

One coal equivalent possesses energy represented by M × 14,000 × 772. Hence we can calculate that one coal equivalent would supply the solar radiation at its present rate for about 2,800 years.

II. The Conservation of Moment of Momentum

We give here an elementary investigation of the fundamental dynamical principle which has been of such importance throughout this volume.

§ 8. Case Where There Are No Forces

Newton's first law of motion tells us that a particle in motion if unacted upon by force, will move continuously in a straight line without change of velocity.

Let A_0, Figure 60, be the position of the particle at any moment. Let A_1 be its position after the time *t*; A_2 be the position at the time 2*t*; A_3 be the position at the time 3*t*, and so on.

Then the first law of motion tells us that the distances A_0 A_1, A_1 A_2, A_2 A_3, A_3 A_4, must form parts of the same straight line and must be all equal.

If lines O A_0, O A_1, O A_2, etc., be drawn from any fixed point 0, then the areas of the triangles O A_0 A_1, O A_1 A_2, O A_2 A_3, 0 A_3 A_4, will be all equal. For each area is one-half the product of the base of the triangle into the perpendicular O T from O on A_0 A_1, and, as the bases of all the triangles are equal, it follows that their areas are equal.

Thus we learn that a particle moving without the action of force will describe around any fixed point O equal areas in equal times.

The product of the mass of the particle and its velocity is termed the momentum. If the momentum be multiplied by O T the product is termed the moment of momentum around O. We have in this case the simplest example of the important principle known as the conservation of moment of momentum.

The moment of momentum of a system of particles moving in a plane is defined to be the excess of the sum of the moments of momentum of those particles which tend round O in one direction, over the sum of the moments of momentum of those particles which tend round O in the opposite direction.

If we deem those moments in one direction round O as positive, and those in the other direction as negative, then we may say that the moment of momentum of a system of particles moving in a plane is the algebraical sum of the several moments of momentum of each of the particles.

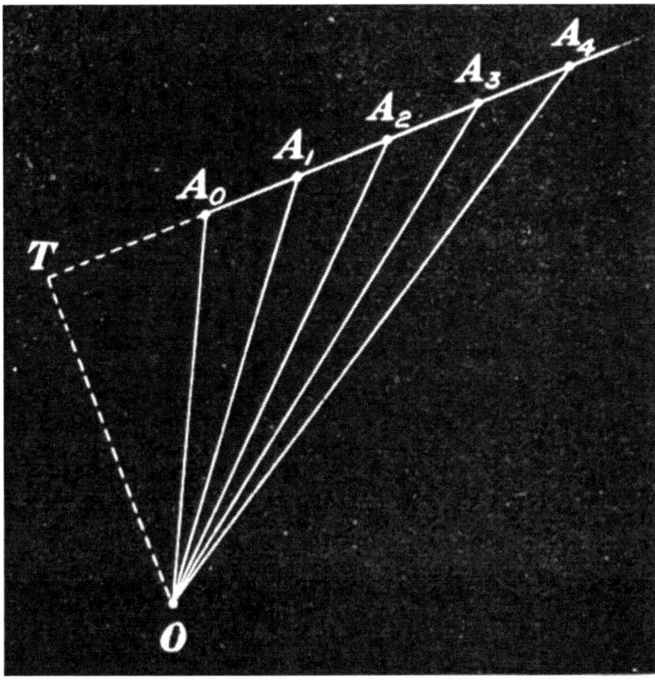

Figure 63. First Law of Motion exemplifies Constant Moment of Momentum.

§ 9. A Geometrical Proposition

The following theorem in elementary geometry will be required:—

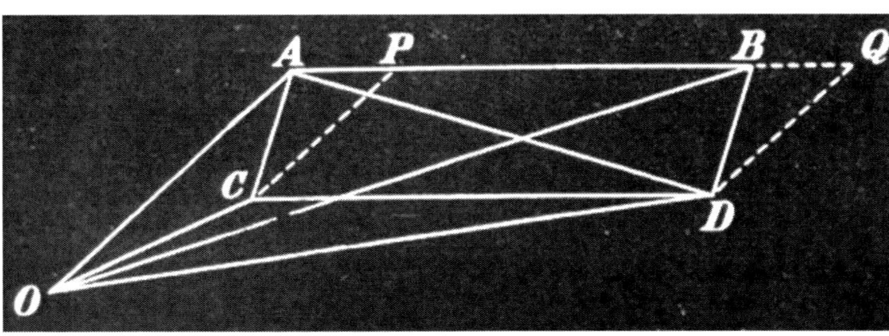

Figure 64. A Useful Geometrical Proposition.

Let A B and A C be adjacent sides of a parallelogram, of which A D is the diagonal, and let O be any point in its plane. Then the area O A C is the difference of the areas O A D and O A B.

Draw D Q and C P parallel to O A. Then O A D = O A Q, whence O A D – O A B = O B Q = O A P = O A C.

§ 10. Relation between the Change of Moment of Momentum and the Force Acting on the Particle

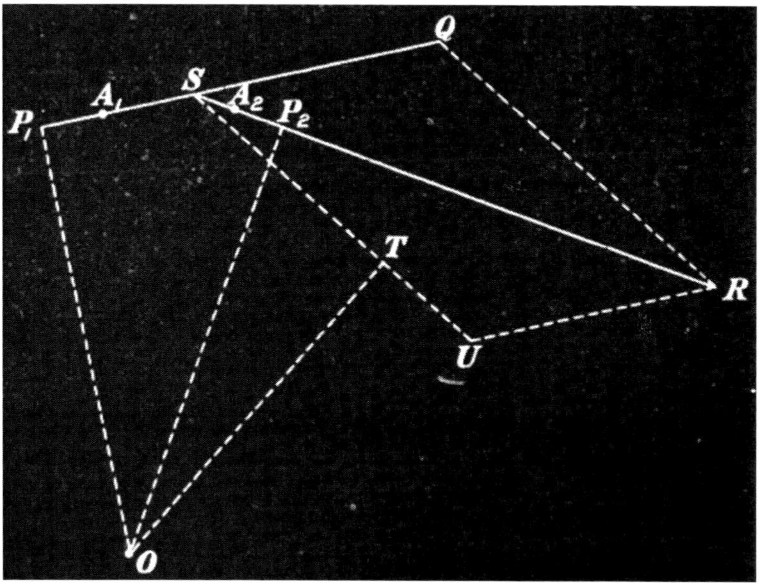

Figure 65. Acceleration of Moment of Momentum equals Moment of Force.

Let A_1 and A_2, Figure 65, be two adjacent points on the path of the particle, and let A_1 Q and A_2 R be the tangents at those points. Let S Q represent the velocity of the particle at A_1, and SR the velocity of the particle at A_2. Then Q R represents both in magnitude and direction the change in velocity due to the force F, which we suppose constant both in magnitude and direction, while the particle moves from A_1 to A_2 in the small time t; we have also Q R = F $t \div m$.

Complete the parallelogram S Q R U, and let fall O P_1, O P_2, O T perpendiculars from O on S Q, S R, S U respectively. Since S Q is the velocity of the particle when at A_1 the moment of momentum is m O P_1 × S Q; when the particle is at A_2 the moment of momentum is m O P_2 × S R. Whence the difference of the moments of momentum at A_1 and A_2 is m (O P_2 × S R - O P_1 × S Q) = 2 m (O S R - O S Q) = 2 m O S U = m O T × S U = m O T. Q R = F t × O T. But in the limit S coincides with A_1 and A_2, and we see that the gain in moment of momentum is t times the moment of the force around O. Hence we deduce the following fundamental theorem, in which, by the expression acceleration of moment of momentum, we mean the rate at which the moment of momentum increases:—

> If a particle under the action of force describes a plane orbit, then the acceleration of the moment of momentum around any point in the plane is equal to the moment of the force around the point.

If the force is constantly directed to a fixed point, then the moment of the force about this point is always zero. Hence the acceleration of the moment of momentum around this point is zero, and the moment of momentum is constant. Thus we have Kepler's law of the description of equal areas in equal times, and we learn that the velocity is inversely proportional to the perpendicular on the tangent.

§ 11. If Two or More Forces Act on a Point, then the Acceleration of the Moment of Momentum, due to the Resultant of These Forces, Is Equal to the Algebraic Sum of the Moments of Momentum due to the Action of the Several Components

Let A D, be a force, and A C and A B its two components. Then, since O A D = O A B + O A C, we see that the moment of A D around O is equal to the sum of the moments of its components. Hence we easily infer that if a force be resolved into several components the moment of that force around a point is equal to the algebraical sum of the moments of its several components.

The acceleration of the moment of momentum around O, due to the resultant of a number of forces, is equal to the moment of that resultant around O. But, as we have just shown, this is equal to the sum of the moments of the separate forces, and hence the theorem is proved.

§ 12. If Any Number of Particles Be Moving in a Plane, and if They Are Not Subjected to Any Forces Save Those Which Arise from Their Mutual Actions, Then the Algebraic Sum of Their Moments of Momentum Round Any Point Is Constant

This important theorem is deduced from the fact stated in the third law of motion, that action and reaction are equal and opposite. Let us take any two particles; then, the acceleration of the moment of momentum of one of them, A, by the action of the other, B, will be the moment of the force between them. The acceleration of the moment of momentum of B by the action of A will be the same moment, but with an opposite sign. Hence the total acceleration of the moment of momentum of the system by the mutual action of A and B is zero. In like manner we dispose of every other pair of actions, and thus, as the total acceleration of the moment of momentum is zero, it follows that the moment of momentum of the system itself must be constant.

This fundamental principle is also known as the doctrine of the conservation of areas. It may be stated in the following manner:—

> If a system of particles are moving in a plane under the influence of their mutual actions only, the algebraic sum of the areas swept out around a point, each multiplied by the mass of the particle, is directly proportional to the time.

§ 13. If a Particle of Mass *M*, Is Moving in Space under the Action of Any Force F, Then the Projection of That Particle on Any Fixed Plane Will Move as if It Were a Particle of Mass *M* Acted upon by That Component of F Which Is Parallel to the Plane

This is evident from the consideration that the acceleration of the particle parallel to the plane must be proportional to this component of F.

Let us now suppose a system of particles moving in space under their mutual actions. The projections of these particles on a plane will move as if they were the particles themselves subjected to the action of forces which are the projections of the actual forces on the same plane, and as the reactions between any two particles are equal and opposite, the projections of those reactions on the plane are equal and opposite. Hence the proof already given of the constancy of the moments of momentum of a plane system, will apply equally to prove the constancy of the moments of momentum of the projections of the particles on the plane. Hence we have the following important theorem:—

> Let a system of particles be moving in space under the action of forces internal to the system only. Let any plane be taken, and any point in that plane, and let the momentum of each particle be projected into the plane, then the algebraic sum of the moments of these projections around the point is constant.

§ 14. On the Principal Plane of a System

Let us suppose a system of particles moving under the influence of their mutual actions. Let O be any point, and draw any plane L through O. Then the moment of momentum of the system around the point O and projected into the plane L is constant. Let us call it S. If another plane, L´, had been drawn through O, the similar moment with regard to L´ is S´. Thus for each plane through O there will be a corresponding value of S.

We have now to show that one plane can be drawn through O, such that the value of S is greater than it is for any other plane. This is the principal plane of the system.

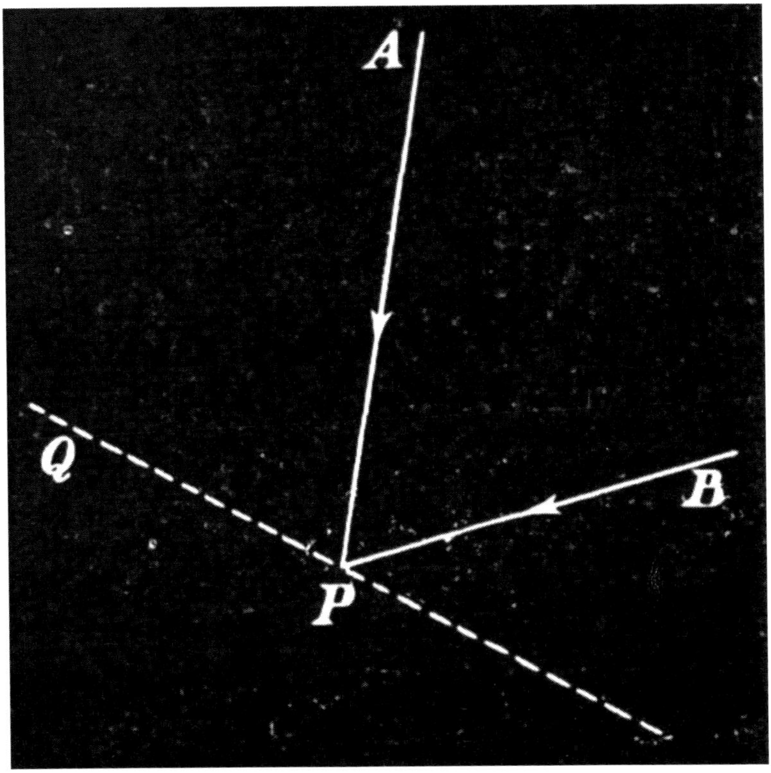

Figure 66. Moment of Momentum unaltered by Collision.

If v be the velocity of a particle, then in a small time t it moves over the distance $v\ t$. If p be the perpendicular from O on the tangent to the motion, then the area of the triangle swept round O in the time t is $\frac{1}{2} p\ v\ t$, and we see that the momentum is proportional to the mass of the particle multiplied into the area swept over in the time t. The quantity S will, therefore, be proportional to the sum of the projections of the areas in L, swept over in the time t, each increased in the proportion of the mass of the particle. It is easily seen that the projection of an area in one plane on another is obtained by multiplying the original area by the cosine of the angle between the two planes. For if the area be divided into thin strips by lines parallel to the line of intersection of the planes, then in the projection of these strips the lengths are unchanged, while the breadths are altered by being multiplied by the cosine of the angle between the two planes. If, therefore, we mark off on the normal to a plane L a length h proportional to any area in that plane, then the projection of this area on any other plane L´ may be measured by the projection of h on the normal to L´.

To determine the moment of momentum resolved in any plane we therefore proceed as follows: Draw a plane through O, and the tangent to the path of one of the particles, and mark off on the normal drawn through O to this plane a length l proportional to the moment of momentum. Repeat the same process for each of the other particles with lengths l', l'', etc., on their several normals. Suppose that l, l', l'' represent forces acting at O, and determine their resultant R. Then R, resolved along any other direction, will give the component of moment of momentum in the plane to which that direction is normal. In any plane which passes through R the component of moment of momentum is zero. The plane perpendicular to R contains the maximum projection of moment of momentum. This is the principal plane of the system which we have seen to be of such importance in connection with the nebular theory.

§ 15. Collisions

The conservation of moment of momentum remains true in a system, even though there may have been actual collisions between the several parts. This is included in the proof already given, for collisions are among the mutual actions referred to. It may, however, be instructive to give a direct proof of a particular case.

Let two particles collide when meeting in the directions A P and B P respectively. Whether the particles be elastic or inelastic is quite immaterial, for in both cases the action and reaction must be equal and opposite, and take place along some line P Q. The action on the particle moving along A P will give to it an acceleration of moment of momentum which is equal to the moment of the action around O. The acceleration of the moment of momentum coming along B P will be equal and opposite. Thus the total acceleration of the moment of momentum is zero. Hence the collision has no effect on the total moment of momentum.

§ 16. Friction and Tides

We have shown that such actions as collisions cannot affect the moment of momentum of the system, neither can it be affected by friction of one body on another. Here, as in the former case, the actions and reactions are equal and opposite, and consequently the accelerations of moment of momentum are zero. Nor is it possible for any tidal action to affect the total moment of momentum of the system. Every such action must be composed of the effects of one particle in the system on another, and as this must invariably produce an equal and opposite reaction the total moment of momentum is unaltered.

INDEX

T

U

V

W